W9-BFZ-880

Technology, Values, and Society

american
university
studies

Series XI
Anthropology and Sociology

Vol. 27

PETER LANG
New York • Washington, D.C./Baltimore • Bern
Frankfurt am Main • Berlin • Brussels • Vienna • Oxford

Mitra Das and Shirley Kolack

Technology, Values, and Society

Social Forces in Technological Change
REVISED EDITION

PETER LANG
New York • Washington, D.C./Baltimore • Bern
Frankfurt am Main • Berlin • Brussels • Vienna • Oxford

Library of Congress Cataloging-in-Publication Data

Das, Mitra.
Technology, values, and society : social forces in technological change /
Mitra Das, Shirley Kolack.—Rev. ed.
p. cm. —(American University studies ; XI. Anthropology and Sociology; v. 27)
Includes bibliographical references and index.
1. Technology—Sociological aspects. I. Kolack, Shirley . II. Title.
T14.5.D28   303.48'3—dc22   2007039743
ISBN 978-1-4331-0189-2
ISSN 0740-0489

Bibliographic information published by **Die Deutsche Bibliothek**.
Die Deutsche Bibliothek lists this publication in the "Deutsche
Nationalbibliografie"; detailed bibliographic data is available
on the Internet at http://dnb.ddb.de/.

Cover design by Clear Point Designs

The paper in this book meets the guidelines for permanence and durability
of the Committee on Production Guidelines for Book Longevity
of the Council of Library Resources.

© 2008 Peter Lang Publishing, Inc., New York
29 Broadway, 18th floor, New York, NY 10006
www.peterlang.com

All rights reserved.
Reprint or reproduction, even partially, in all forms such as microfilm,
xerography, microfiche, microcard, and offset strictly prohibited.

Printed in the United States of America

To
Mukti and Abhijit
and
Sol, Joshua, David and Michael

Preface to the Second Edition

The first edition of this book was published in the last century. The 21st century builds upon the trends and forces started in the century before it. People around the world living in many different regions of the globe have experienced dramatic changes in their personal and social lives. Some of these changes have been due to the introduction of new technologies that have opened up new possibilities of economic opportunities and outcomes. These technologies have been associated with and prompted the emergence of social and economic forces to accommodate them. An example of this is the Computer technology which has connected the world into a global economy.

In this edition we have included a new chapter on the Computer Revolution using the case of Israel. By so doing we have attempted to expand the theme that we elaborated in the first edition of the inter-dependency of technology, society and values and show how each of these factors affect the others. The Computer Revolution is affecting societies around the world in myriad ways. Not only are relations among members of the society impacted; relations between nations are also affected. The Computer Revolution can be seen as a new mode of production which organizes and replaces human labor just as the industrial system did when it first surfaced in areas it entered where people principally eked out their living from the soil or resources naturally available in their environment.

In writing the chapter on the Computer Revolution we have received assistance from Kimberly Leiken who has generously given her time and energy in completing this work. She has also helped us with the editorial work in this revised edition. We gratefully acknowledge her assistance. Abhijit Das has helped us with the maps in this revised edition and we gratefully acknowledge his assistance. We also want to thank Bernadette Shade at Peter Lang for her prompt help in completing this edition. For a long time we have been thinking of expanding our original work by adding a chapter on computer technology which has led to far reaching structural changes in societies that have accepted them. We are indeed very grateful to Chris Myers of Peter Lang to have this opportunity to do it now.

Mitra Das & Shirley Kolack
Lowell, Massachusetts
2007

Preface to the First Edition

The germ of the idea for examining specific case studies to show the interrelationship of technology, values and society came from our teaching in a program to develop new curriculum in the area of technology and values sponsored by the National Endowment for the Humanities at the University of Massachusetts Lowell. We found that for the subject matter to have meaning our students needed examples of different types of societies—primitive, developing and industrial—where there were interactions between technological and social developments.

We have sought to demonstrate how technological change can affect the social and personal lives of members of society, and how the social structure of a society, in turn, can foster or obstruct technological change. We hope that what the reader will learn from this book will affect the way in which he looks at his life and society.

Many colleagues and students helped us along the way. We especially wish to thank Barbara Miliaras for generously giving of her time in skillfully copy editing our manuscript. We are grateful to Melvin Cherno and Judith Pastore for reading and commenting on earlier versions. Sharon Quigley, Stella Klesaris and Pat Dailey graciously and efficiently typed and word-processed the manuscript.

Peter Blewett, Acting Dean of the College of Liberal Arts, supported and encouraged us. Kazi Belal helped with reproduction of photographs. Barbara Boucher and Swapna Das graciously drew the illustrations. Cyrus Mehta provided access to library resources essential for our research. We are sincerely grateful to all of them.

Mitra Das & Shirley Kolack
Lowell, Massachusetts
June 1988

Table of Contents

CHAPTER 1

Introduction:
Technology and Society

This analysis of the relationship of technology and society will demonstrate the dynamic interaction that exists among values, technology and society. This phenomenon becomes evident when the technological bases of societies, both preliterate and literate, traditional and modern, are examined. An exploration of the tools used for production by a people reveals the society's level of technical skills and scientific development; more importantly, it sheds light on its economy and social organization.[1]

Technological innovation affects the lives of people in myriad ways. This becomes evident when the change relates to a single new item introduced into the culture and particularly when the entire society is in the process of moving from one mode of subsistence to another. These relationships are explored by an examination of different societies at different levels of technological development.

The emergence of a primitive hunting technology involving simple tools was the first great technological breakthrough. This was followed by agricultural developments that led to plant cultivation, which had far-reaching social consequences, since now food could be stored and replenished. Thus, population size was no longer partially controlled by the lack of food resources. Because hunting and gathering societies required physical mobility, it was dysfunctional to have large numbers of children to take along in the search for food. As a result of the agricultural revolution, however, agriculturists, living in settled communities, found additional children beneficial in helping with chores. Moreover, some members of agricultural societies were now free to engage in pursuits other than food gathering, resulting in a more elaborate social structure with a division of labor that allowed for occupational specialization.

While it is true that technological changes affect social life, this relationship is not one-sided. Societies do not passively accommodate or receive new technology. Instead, through their value premises and structural arrangements, they promote or inhibit technological development. There is an intricate interrelatedness between technological and human factors.

Thus, industrialization in Lowell, Massachusetts, was inevitably linked with the ideological and political revolution taking place in nineteenth century America. Technology came to be viewed not only as a means of promoting material welfare and progress, but as an ingredient for American democracy in a society predicated on values of liberty, virtue and independence. Personal discipline and restraint were required of factory workers, characteristics seen as beneficial to human conduct. Indeed, the introduction of the factory system and labor-saving machines was considered indispensable for liberty and freedom. The promise of labor-saving devices strongly appealed to a nation eager for economic independence, the safeguarding of moral purity and the promotion of industry and thrift.[2]

The characteristic qualities of mechanization—regularity, uniformity, subordination, harmony and efficiency-appeared to offer a model for government and society in general. During the American Revolution, the view was expressed that republicanism would not be secure until freedom from the corrupting influences and dependence upon European manufacturers and trade was achieved. Men who upheld a conservative ideology of republicanism began to glorify machines not simply as functional but as signs and symbols of the future of the nation.[3] Ideology alone did not cause the industrial revolution in Lowell, but it did shape the basis for action. The desire for industrial progress went hand in hand with providing the proper institutional environment.[4]

Furthermore, industrialization in Lowell, Massachusetts, was designed to take place in a planned work setting that would superimpose the cherished values of agrarianism on the conditions of manufacturing. Remaining within an essentially agricultural context instead of clustered in the cities, manufacturing would enhance rather than threaten the independence and virtue of American society. Technology would be a boon to social order and control.[5] However, the social organization of the factory system and the technological patterns relating to labor-saving machines and cost-cutting became linked to profits; in such a milieu, the old agrarian values could not survive. A market economy inevitably established its own scale of values.

In any technology, the techniques and practical know-how involved in a society's capacity to produce goods and services will have an effect on the network of social relationships in work and in all other institutions. The values, the standards that uphold the norms of a society, will also play a part in determining how the productive forces of a society are organized. When a new technology is introduced, there is often a value lag, for new technologies require new social relationships.[6] Accelerated technological change often brings disruption.

The formula that ideology follows technology or that technology follows ideology is too simple. This relationship is rather a dialectical interchange in which both are transformed. The case studies analyzed in the chapters that follow shed light on the intricate connections among technology, society and values. The effects of techology and values on societal institutions are explored. That different forms of technology used for production have affected social relationships and social organization is documented. Technologies that are socially significant and have led humankind on an irreversible path of increased division of labor and other forms of social differentiation are probed. Hence, hunting, agricultural and industrial societies provide the focal points for analysis.

The earliest mode of man's subsistence was hunting and foraging. Hunting and foraging made status differences based on sharp inequalities of wealth impossible. The means of production in the form of tools and resources were available to all. All members had to share in order to exist. Agriculture and industrialization led to control of the means of production in the form of land and machinery in the hands of a few. The emergence of agriculture, aptly labeled a revolution, altered the social relations of man to man, and man to nature. It ushered in forms of hierarchical relations between people based upon the relationship of an individual or group to land and the surplus produced. Similarly, industrialization—another revolution experienced by mankind—further altered the existing structure of social relationships. It led to greater internal differentiation, which included status and power differences, as well as a vast division of labor among its participants. The legacy of industrialization was an expansion of material productivity unprecedented in human history. It also enabled people to live longer.

In this text, each of the three types of societies highlighted, hunting, agricultural and industrial, has a distinct form of technology. The effect of technology on each society and how the society is affected by the sustaining social base is demonstrated. A recurring theme is the interdependence of technology and society. Neither is ever autonomous. Technologies are not neutral, nor do they always spell unencumbered progress. The tenet of technological determinism is challenged. The impact of a society's values on the shape of its technology is scrutinized in an effort to understand the ways in which technologies become intertwined with human existence. How human activities influence the development of technologies is also examined.

Hunters and foragers are exemplified by the Mbuti Pygmies of Zaire and the Alaskan Eskimos. The Pygmies hunt on land with bow and arrows; some Eskimos hunt sea animals with harpoons. Two types of agricultural systems are investigated: one cultivates land with the hoe and digging stick; the other uses the plow. East African villages in Kenya demonstrate the hoe and digging-stick agricultural systems. Hoe and digging-stick tools are contrasted with plow technology. Villages in the Punjab region of India provide the case studies of plow cultivation. All of these case examples are compared with an industrial society: Lowell, Massachusetts, the birthplace of the industrial revolution in the United States. The patterns of work associated with each technology are strikingly different from one another. Each of these technological developments, however, has altered the character of work in their individual time and place.

The text further uncovers the anticipated as well as the unanticipated consequences of the modes of technology utilized by each society, exploring technology as a mechanism for social change. When a society changes its technology, unintended repercussions are likely to be felt in a broad range of customs and institutions. When primitive people turn from a life based on hunting to one based on planting crops and raising cattle, their family, government and religion undergo a complete transformation.

Hunting and gathering societies are cooperative and sharing. When the rifle—a new technology—was adopted for use by Eskimos, there was less need for tribes to cooperate and share; each hunter could operate alone. The introduction of a cash income eroded cooperative efforts even further and brought about new demands and desires for consumer items. Cash income was provided by such new economic activities of the Eskimos as the trapping of animals for fur by individual entrepreneurs.

For the Mbuti Pygmies, shared economic activities were essential to survival. The life of the Pygmies of the Congo revolved around cooperative hunting. The forest was the source of their food. After hunting, they had time for relaxation. In recent times they have been encouraged to come out of the forest and abandon their mode of existence.

The Pygmies have come to rely on more sophisticated tools, such as metal-tipped spears supplied by the tribal villagers who live in settlements outside the rain forest. Forsaking their traditional life-style based upon hunting will no doubt lead to an increased dependence on technology alien to their culture and will affect their symbiotic relationship with the natural environment. The way will undoubtedly be opened for less equality in economic relationships. Hunters generally live in small migratory bands, lack property, are highly democratic and tend to be monogamous. By contrast, those engaged in agriculture live in larger and more permanent settlements; accumulate land, have powerful leaders and husbands often have several wives. Hunters and gatherers are more willing to take risks and are entrepreneurial. Agriculturists are more apt to be conservative, protect their land and resist change.[7]

The development of agriculture resulted in an increased work load per capita, since farming is more labor-intensive than hunting and gathering. In Indian villages, where the plow is the primary agricultural tool, women are often removed from the agricultural process and men are put exclusively in charge of farming. Consequently, the position of women is subordinate to that of men.[8] When the plow and draft animals replaced hand tools, the role of women in the agricultural process declined. In contrast, in many African villages based on hoe technology, the farming activities of sowing, weeding and harvesting are primarily the chores of women. These women produce food for their families and even cash crops using the primitive methods of hoe cultivation.

African villages have remained more egalitarian than Indian villages. An often unanticipated consequence of the change to a more labor-intensive land use and its accompanying more sophisticated form of technology is a downgrading in the status of women and a more generalized system of status inequality. Where women are part of the cultivation process, husbands often pay a bride price. Where women are not a part of this process, wives often pay a husband price.[9]

The shift from animal and human resources to machinery as the main source of power during industrialization also brought profound changes in the nature of the labor force, the conditions and definition of work, the development of an elaborate class system and even a new political ideology represented by industrial capitalism. Through the introduction of machine technology to Lowell, a farm community over time was transformed into a grimy factory town.

The initial work force in Lowell consisted of local farm women, but, in time, conditions worsened; new immigrants began to dominate the work force and profit supplanted all other considerations. Lowell provides a dramatic example of a community that experienced drastic changes as it moved from one mode of production to another. A brief period of growth and prosperity in Lowell was followed by a precipitous economic decline. It is only recently that a revival of economic activities has taken place as a computer-based technology developed in the area.

An examination of different types of societies markedly different in terms of technology and social organization will aid in understanding the ways in which societies react and adapt to the onslaught of technological advancement. Cross-cultural comparisons based on empirical evidence are offered to demonstrate that neither technology nor values are ever autonomous or always spell progress.

The paradigm provided for analysis will have application to both preliterate and modern societies. The significance and relevance of the cases studied lie in their utilization as models for understanding the intimate and dynamic connections among technology, society and values in ways extending beyond particular locales. The interactive effects of technology and values on all of a society's social systems—family, economy, religion and ideology—is demonstrated. That all major institutions reflect the impact of the interplay between technological and social forces is illustrated. Extensive cultural changes are brought about by advances in agriculture and by the introduction of the factory system and mass production.[10]

The variables influencing how a society adjusts to the onslaught of technological change are uncovered. Material culture generally outpaces the development of nonmaterial culture and thus there is often cultural lag.[11] For a period of time, values remain in place that are inappropriate to technological advancements. This often leads to discontinuity and disorganization, since social relationships that are not useful for new productive forces retard progress. The social network of the family economy in Lowell was not functional for the productive processes required for industrialization. A new form of work organization, similar to the shift from hunting and gathering to agricultural social structures, was demanded. In Lowell, the increased emphasis on individualism as expressed in the values of the nuclear family did provide the motivation for the young, unmarried farm women to come to work in the mills. A desire for mobility as expressed in new family norms was also an impetus for the women to leave the farms and for their willingness to take up a new mode of life in the city.

Value premises often provide the thrust for technological innovations. The emphasis on values associated with work, the desire for progress and greater productivity, led men such as Francis Cabot Lowell to develop power-driven machines that were far more productive than human or animal labor. Such technology in the form of a novel source of energy became a prime mover in the modern stage of cultural development. Machines began to dominate human activities and change the nature and meaning of work.

The introduction of the factory system brought profound changes in work organization. A detailed division of labor enhanced the role of semiskilled workers at the expense of the multiskilled craftsmen. There was a centralized, fixed workplace, the imposition of a disciplined work schedule and a strict workweek. All of this led to vertical stratification in the workplace and thus to more control over workers. This was in contrast to the putting-out system of the family economy where the capitalist distributed raw materials on a piecework basis to workers for production in their own homes. Within the manufacturing establishments, no one person ever produced a total marketable product.

A society's values may provide a powerful impetus for either adapting to technological innovations or to maintaining the status quo. For centuries, the Indian agricultural system resisted technological innovation because the social system inhibited its development. Technology remained stagnant. Agricultural implements and techniques—crops were not rotated—did not change significantly between the time of the Moguls and the early twentieth century. Social arrangements on the land played an important part in this lack of progress. The Indian peasant was seen by the ruling classes mainly as a producer of revenue. The more the peasant produced, the more he had to turn over to the tax collector. Therefore, the small farmer had no incentive to produce more. In addition, the Indian caste system promoted a docile labor force. There was an exchange of labor and services for food between the higher castes who had land and the lower castes who had little or none. The higher castes often preferred smaller returns on their land in exchange for less supervision over their workers in order to compel them to improve their productivity.

People are likely to be receptive to technological innovation only when it is perceived to be in their interest and serves their purpose.[12] At any period in time, a society operates under a set of social conditions and a world view that influences its receptivity to change. Its desire to accept or change these forces is linked to its value underpinnings and its world view. No technology can be sufficient in itself. To succeed it must solicit more effective social arrangements for achieving greater productivity. A shift in the organization of people around their work is always mandated by a change to a new technology.

At the present time, technological forces are affecting and in turn affected by man's social situations worldwide. It is crucial that these forces be revealed. The interrelationship of sociotechnological systems must be understood in the postindustrialized era of computer technology. The role of technology as an expedient in shifting cultural customs, norms and traditions must be fathomed. When a people change their technology, unintended consequences are felt in a wide range of values and institutions. The response of societies to the irresistible presence of technology in the form of new energy sources, work organization and its accompanying assault on social institutions is elaborated in the chapters that follow.

Sustained technological growth is a comparatively recent phenomenon. Most hunting societies persisted in their traditional ways and only a few made the leap into agricultural technology. There were, in turn, many stages in the development of agricultural technology, beginning with simple garden tools and culminating with large-scale farming, domesticated beasts of burden, heavy plows, crop rotation, irrigation and flood control. The pathway from a low agricultural technology to a higher stage similarly was taken by very few societies.[13] In turn, only a few of the agriculturally successful societies were able or disposed to take the next step into industrial technology. Apparently, only in industrialized societies have there been institutions, resources and wealth committed to economic growth, mass production and systematic rational planning. Also, only with the coming of industrialization did science and technology become systematically interlocked. Technology is not self-perpetuating; whether it develops and grows depends on the institutional framework as well as the cultural and intellectual milieu.

Characteristics of technology in different types of societies and the sources, directions and implications of technological progress are scrutinized. An interpretation of the developments of historically significant technological changes is provided by examining case studies of varying complexity. The impact of simpler technologies is most easily understood. Therefore, the analysis proceeds from a discussion of simple to more complex societies. The reader should learn in the chapters that follow more about the nature and extent of technology's influence on social institutions and individual lives. Also probed is when this impact is beneficial and when it is harmful. Technology is a vehicle for both alienation and liberation.

Technology does permeate every corner of our lives—overwhelmingly shaping and transforming them for better or for worse. However, technology, though dynamic, cannot act independently of values. Some of the value influences explored are attitudes toward domination of nature; which groups in society will control technology; and whose interests will benefit from its use.

Today, an incredibly powerful new computer technology has emerged. Yet, underlying continuities link this new technology with the simpler technologies of the past. Computer technology, as did the other technologies that it supplanted, is profoundly influencing the world in which we live in ways that are not completely foreseen. An examination of the technological, societal and value interactions in the societies we explore will aid in comprehending the role of this new electronic technology in contemporary society. We will be in a position to appraise more critically the implications of technological advancement and its inevitable economic, social and cultural consequences.

Technological change has major cultural ramifications. This book explores the nature of these social, economic, and political changes brought about by shifting technology. *Technology, Values, and Society* discusses the major technological advancements throughout history, and provides analysis of the cultural alterations brought about as a result. The final chapter looks at the computer revolution and it's effect on men and women all over the world. Using Israel as a case study, we are exploring the ways in which the computer has changed our lives.

Technology is strongly intertwined with our future in ways we do not fully comprehend. In particular, all of the effects of the recent Computer Revolution are yet to be discovered. It has certainly opened an immense range of prospects to a variety of people, including buying and selling products online, decreasing the time and effort necessary to complete certain tasks, and creating a new assortment of employment opportunities. Yet, there are many negative traits of this development which involve the loss of many other employment positions, being constantly connected to one another at all times, and internet addiction and abuse. While globalization has created a greater bond among people and countries, it remains to be determined whether or not this is beneficial to the modern world. However, we should critically use the knowledge that is available in assessing its impact on the shape of societies. It is our hope that what we have written will aid in this process.

Notes

1. Karl Marx, *Contributions to the Critique of Political Economy*, Cambridge, Mass.: Harvard University Press, 1970.

2. For an elaboration of these relationships see John Kasson, *Civilizing the Machine*, New York: Penguin Books, 1976, pp. 3-51, 55-106.

3. Kasson, p. 62.

4. Kasson, pp. 68-69.

5. Kasson, pp. 3-51.

6. William Ogburn, *Our Cultural and Social Changes: Selected Papers*, Chicago: Chicago University Press, 1964.

7. Richard A. Barrett, *Culture and Conduct*, Belmont, Calif.: Wadsworth Publishing Company, 1984, pp. 84-85.

8. Ester Boserup, *Women's Role in Economic Development*, New York: St. Martin's Press, 1970, pp. 50-51.

9. Boserup, pp. 50-51.

10. Marvin Harris, *America Now*, New York: Simon & Schuster, 1981, p. 12.

11. Ogburn, 1964.

12. Barrington Moore, Jr., *Social Origins of Dictatorship and Democracy*, Boston: Beacon Press, 1966, pp. 314-385.

13. Bernard Gendron, *Technology and the Human Condition*, New York: St. Martin's Press, 1977, p. 26.

CHAPTER 2

Hunters and Foragers
of the Rain Forest: Pygmies

The Physical Setting

The Ituri Forest, the home of the Mbuti Pygmies, is in the heart of Africa near the banks of the tributaries of the Congo River. One of the least-known regions of the world, it lies a few miles north of the equator. It consists of a zone of deep equatorial forest that covers two-fifths of Democratic Republic of the Congo (DRC). It is this Africa that travelers and explorers, awed by its beauty and mystery, called the Dark Continent. This forest is a natural botanical wonderland where creatures large and small, human and non-human, have created a niche for themselves. It is a forest that has persisted for unknown millennia, untouched and unchanged until recently. The Ituri Pygmies, known as Mbuti, have lived here for centuries. The earliest written record of their existence dates back to 2250 B.C.[1]

It is difficult to ascertain how long the Pygmies have inhabited the rain forest. They are reported to be a prehistoric group of hunters who have survived into the twentieth century.[2] In fact, they may be the original inhabitants of the primeval rain forest. They are known to have inhabited the entire belt of the rain forest, which spreads across the equatorial stretch of the African continent. It is estimated that the Pygmies occupied the land "stretching from the west coast right across to the open savanna country of the east, on the far side of the chain of lakes that divided the Congo from East Africa."[3] There are two major Pygmy concentrations: one is located northeast of the mouth of the Congo River in modern Gabon and Brazzaville; the other is

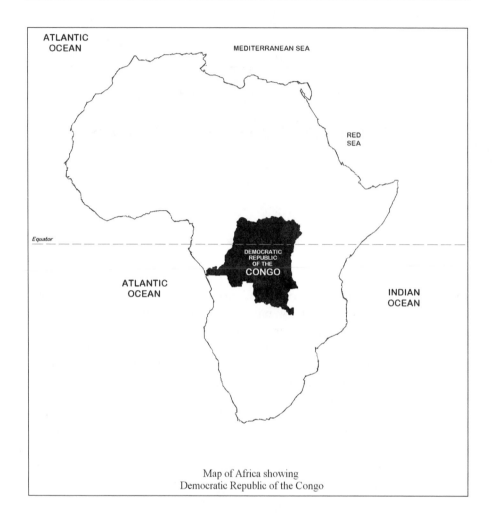

Map of Africa showing
Democratic Republic of the Congo

located in the Ituri Forest northeast of the intersection of the Lualba and Lomani rivers.[4] Our discussion is limited to the Mbuti Pygmies who inhabit the Ituri rain forest.

The rain forest covers most of the northern half of Democratic Republic of the Congo including much of the Congo River Basin. The forests are now shrinking as trees on their borders are cut to make room for plantations for the nearby tribal villagers. These inroads into the forests affect the life-style of the Pygmies who live close to the village settlements. As a result of this cultural

contact with the tribal villagers, many Pygmy hunters abandon their traditional way of life in exchange for more sedentary ways of living. This, of
course, affects them in serious ways, robbing them of the autonomy and
equality that has been the mark of their culture. Other Pygmies living in the
interior of the forest, however, still continue their accustomed hunting and
foraging patterns of subsistence.

The equatorial rain forest covers a vast expanse of terrain and natural
vegetation, thick, dense and rich; at points it is almost impenetrable. Sunlight
reaching down filters through a network of leaves growing on gigantic trees.
The interlaced tall trees form a canopy that provides a natural screen shielding
the forest from the hot, glaring sun of the equator. The vegetation is so dense
that great distances are obscured. Two major patterns of natural vegetation are
discernible in the rain forest. Much of the forest consists of swamp and a still
larger area is a mixture of marshy and firm land. The rich variegated flora
sustained by the rain forest, nourished by a regular supply of rainfall, provides
a damp, cool habitat. The trees that remain unexposed to direct sunlight never
completely dry.

Although it was the middle of August in 1946 when Anne Putnam, the
wife of an American anthropologist Patrick Putnam, first visited the Ituri
Forest, she was pleasantly surprised by its coolness. The cabin that was to be
her home for the next few years had a fire burning within it, even though it
was situated on land on the equator.[5]

The variegated fauna and flora of the rain forest provide sustenance for
the Pygmies. They live by hunting and foraging. This is the earliest mode of
human subsistence. Hunters and gatherers, in contrast to cultivators, do not
produce the food they consume. They depend upon what is available in the
natural setting. It is only when the environment fails to sustain them due to
depletion of resources that they turn to cultivation.[6]

The Mbuti Pygmies managed for centuries to exclude the outsiders from
the forest. The outsiders remained near the perimeter of the forest. Even
Belgian colonial administrators did not disturb the tranquility of the forest
and its people. However, the seeds for change that affected the Mbuti were
sown when the colonial government imposed taxes and demanded that the
village farmers living around the forest grow cotton.[7]

The rain forest, replete with rich fauna and flora, allowed the Pygmies
since time immemorial to hunt and forage for their food. Indeed, the thick
and dense forest may have inhibited the development of other modes of

subsistence by making its perimeters unapproachable to outsiders. While the riverbanks provided limited possibilities for crop cultivation, the core of the forest was inhospitable to such developments. The Pygmies remained self-sufficient. For unknown millennia they were unaffected and isolated from the rest of the world. Biologically and culturally, Pygmies adapted to the forest, maintaining equilibrium between them and their environment. They were able to continue thus, because of the favorable ecological conditions surrounding them, which included proximity to water, rich vegetation, abundant supply of game and moderate climate. These conditions are prerequisites to sustaining hunting and foraging societies.[8]

The Pygmies are faced with forces that threaten to disturb the balance that had evolved between them and nature. Pygmies are now in a state of transition, and soon their fascinating culture, an illustrious example of the oldest form of human adaptation since we emerged as cultural beings, may become a thing of the past.

Ethnic Background

The Ituri Pygmies are racially and culturally distinct from tribal villagers—Bantu and Sudanic people—who live around them. They have unique physical features that make them distinct from all other peoples of the world. Their most unique physical quality is their height. Men attain an average height of 4'8 1/2" and women 4'6". The skin color of the Pygmies ranges from yellow-tan to red-brown. They are relatively fair-skinned and, unlike the tribal villagers, do not belong to the contemporary world of Black Africa.[9] The Pygmies have protocaucasoid traits such as thin, uneverted lips, prominent eyebrow ridges and a very heavy growth of beard and body hair. Culturally, the Pygmies as a unit are distinct from the tribal neighbors. They are hunters and gatherers, not cultivators like the tribal villagers. Thus, their social structure is distinct.[10] It is egalitarian.

The Pygmies appear to be divided into distinct linguistic groups. Each Pygmy group has apparently adopted the languages of the different African immigrants who settled on the outskirts of the forest and with whom they came into contact. In spite of this linguistic diversity, the Pygmies of the Ituri Forest recognize their common identity. This is because their intonations and speech accents are strikingly similar and are distinct from the African intonations.[11]

The Pygmies are most at home in the forest. Yet they live in two worlds, their own and that of the villagers. They trade with the villagers the products of the forest in exchange for such items as food and metalware produced by the villagers. This relationship is not purely economic. It is religious and political also. The Pygmies participate in village ceremonies and rituals and the villagers refer to the Pygmies as their servants. However, this relationship is not stable or permanent and is broken when the Pygmies so choose it.

Technology of Foraging and Hunting

The Pygmies have developed "forest skills" that allow them to obtain food from the forest. Like hunters elsewhere, they do not produce their food. They rely principally upon undomesticated food found in nature. Their skills have made them so adept in the forest that they think of themselves as "the people of the forest." Outsiders also confer this title on Pygmies. Their knowledge of the forest flora allows the Pygmies to survive and satisfy their food requirements. It has also aided travelers and explorers who otherwise would have perished without the Pygmies' aid in identifying edible wild roots and beans that grow so profusely in the forest. For example, Stanley, the explorer, famous for his work *In Darkest Africa*, first met Pygmies in the Ituri Forest in 1887. During his expedition, he and his men would have died of starvation had it not been for the Pygmies, who taught them how to survive in the wilderness. These skills of the Pygmies, developed from living in the forest, also permit them to engage in economic exchange with the surrounding villagers with whom they trade their foods in exchange for items unavailable in the forest.

The technology of the Pygmies is pre-Stone Age. It is simple and rudimentary and has remained so for centuries. Pygmies do not make tools from any other material than what is found in the forest. These include wooden arrows and spears, hunting nets and baskets. The arrows and spears are made from dead branches and twigs that are sharpened and fire-hardened at the tips. The nets used to trap animals are made of bark of liana and kusa vines, laced into a mesh to form yards of hunting nets the size of tennis nets. Split bamboo is used as a cutting utensil, and a digging stick to extract honey.

The Pygmies have not learned to fashion tools out of stone or iron. However, metal weapons and utensils have entered the Pygmy economy. These have been introduced by the neighboring villagers who live at the borders of the forest. Metal knives and spears are exchanged by the villagers for the prized meat of the forest. The machete and the ax blade needed both for hunting and domestic use are acquired by the Pygmies from the villagers. Even the cooking pots they use for preparing meals come from the sedentary tribal villagers. Pygmies have not learned to make fire; they carry hot embers wrapped in fire-resistant leaves when they move from camp to camp. Technologically, the Pygmies have remained very simple. Yet, with simple and primitive hunting equipment they feed themselves and also provide meat for their cultivating neighbors.

The Pygmies hunt alone as individuals as well as collectively in groups. Their hunting techniques and tools vary with the animals that are stalked. Wooden arrows and metal spears are used in solitary hunting. Birds, monkeys and small animals are killed with arrows tipped with potions made from the saps of forest vines. These potions are poisonous and paralyze the victim. As the animal succumbs to the inflicted wound, it is retrieved by the hunter who follows his prey very carefully so as not to lose it once it falls. Metal-tipped arrows are also used to hunt such animals as forest antelopes and wild pigs.

Elephant stalking is dangerous and involves one or two individuals who follow the tracks of this gigantic beast, pursuing it at a short distance. The hunter will either kill or be killed by the elephant. Metal spears are used to hunt and kill elephants. The individual hunter has to so approach the elephant that he can pierce its underbelly with a spear and inflict upon it a lethal and deadly blow. Once the elephant is killed, the entire band is notified and all arrive at the scene of death to claim their share of meat. There is much joy and rejoicing following a successful elephant hunt.

Net hunting is a group phenomenon and involves the entire Pygmy band in a cooperative endeavor. In fact, it is impossible to hunt alone with nets. Adults of both sexes as well as children participate in this kind of hunt. Only very small children are excluded.

The nets are first set up as traps in the appropriate places, where the men wait. The women and children then shout and beat drums, directing the animals toward the nets. They have to be particularly cautious so as not to forewarn the prey and thus lose their game. As the fleeing animals panic and run toward the hidden nets, they are trapped and then speared by the men. Killed animals are transported back to camp in baskets carried by the women.

In addition to the animals that are hunted for food, plants are used to supplement the Pygmy diet. Edible vines, roots, mushrooms, nuts and berries are collected as food. Honey located in rotting trees, where bees have built their hives, as well as a variety of leaves are considered delicacies. These are collected with bare hands, sometimes with the aid of a digging stick. Pygmy men and women, through observation, experience and generations of forest living, have learned to recognize and distinguish edible from poisonous sources of food.

The Social Organization

Every human society, irrespective of its size, has a social structure or organization. This social organization is affected by its mode of subsistence. This is certainly true of the hunters and foragers represented by the Pygmies of the Ituri Forest.

Pygmies share features common to all hunters. Hunting imposes a certain degree of nomadism and splits up large populations into smaller groups known as bands. Bands fit in with the nomadic life style of hunters, who are required to move according to the availability of food. This mitigates against the acquisition and accumulation of material wealth and has a leveling influence, eliminating distinctions associated with private property.

Pygmies of the Ituri Forest live in bands that vary in size from twenty to one hundred and fifty people. These bands move around throughout the year within the natural boundaries of the forest. Pygmies are nomadic; yet they have a strong sense of territorial permanence. Each band recognizes, however, its own hunting territory within which it normally hunts. It is this territory that the band refers to as its home.[12] Generally, bands do not hunt near territorial boundaries if neighboring bands are in the region. However, boundary crossing is permitted under specific conditions. These include such conditions as the necessity of stalking pursued animals. When pursuit of the hunted animals leads band members into the territory of another band, they are required to share their catch with the band whose territory has been trespassed. The territory within which a band moves has resources sufficient to sustain its needs. A territorial claim only establishes first rights of access to resources. It does not prohibit others from access to the produce of the forest.

A Pygmy band consists of several nuclear families. Each nuclear family is analogous to a single cell; a grouping of these cells comprise the band. These nuclear families within the band may be related by kin ties. Indeed, the presence of kin may attract individual families to join a band. However, composition of a band is not determined only by the principle of kinship through lineage membership. Economic considerations are more important in determining the size and structure of a band. This is evident from the examination of the processes of fusion and fission that constantly characterizes the life of a band.

The Pygmy band, although territorially bounded, is not limited to a fixed group of individuals or families. Its membership boundaries are open and therefore flexible. The band's membership is always in a state of flux. This impermanence is due to personal and economic factors. Individuals unable to get along with other band members may resolve their problem by leaving the band. Others may join a camp, thereby becoming a part of it. Still others, after splintering from the original band, may choose to move alone. Thus, bands may come together and then separate due to personal reasons of hostility and conflict.

The turnover in band membership is also affected by economic factors that influence hunting such as season and time of the year. For example, net hunters are known to split up into smaller groups to maximize their search for honey during the honey season, only to regroup later with the conclusion of this season.[13] Any hostilities between band members are buried by the antagonists joining different bands at the time of regrouping. Bow and arrow users similarly disperse and come together under the same conditions, but for opposite reasons.[14] During the honey season bow and arrow hunters form larger groups, claiming this is the most efficient way of gathering honey. Net hunting bands are large, as required by the necessity of their hunting technique. Bow and arrow hunters need to move in small bands for maximum hunting efficiency. Both kinds of bands, however, divide and realign changing their size and composition as needed. Hence, the band does not remain constant and its membership changes according to the food in season, the pattern of cooperation required and whether hostilities develop among those comprising the band.

Band organization functions well with the nomadic life-style of hunters and foragers. The quest for food compels the Pygmies to move to sites where food is available. It also necessitates cooperative efforts based on the consideration that the social good outweighs the needs of the individuals. This ensures the survival of the group. Whenever a campsite becomes unproductive, it is abandoned. On very short notice, the Pygmies pack up and are ready to move to newer hunting grounds. Mobility and cooperation go hand in hand. Both are indispensable for survival.

The movement and availability of the animal population and the existence and distribution of fruits, vegetables and edible roots according to seasonal variations require that groups subsisting on them be mobile. Nomadism, therefore, is built into hunting and foraging modes of subsistence. It limits population size. The relationship between nomadism and reduced fertility was documented as early as 1922.[15] Mobility is difficult if women are burdened with the weight of several infants whom they have to carry long distances as they move from site to site in search of food or a new residential campsite. Hunters and gatherers have effectively reduced fertility by prolonging lactation. Prolonged lactation prevents ovulation by reducing the energy reserve required by the body to ovulate. The change from nomadism to a sedentary way of life increases the population by producing a situation where increase in the number of children to care for does not adversely affect the work efforts of individual mothers.[16]

Nomadism also restricts the growth of private property. As Pygmies move in the forest to new grounds, they are known to leave personal things such as cooking pots that are too heavy to carry and are not needed for preparation of food in the forest interior. The lack of attachment to material things, necessitated by nomadism, is functional in the forest environment, both physically and socially. It leads to sharing and caring. Mobility thus is based on economic factors and their consequences. It also has political ramifications; as noticed in the changing composition of the band, individuals are able to put an end to intraband hostility by breaking away and joining a new band.

Besides the band, the other most important unit of the Pygmy social organization is the family. Upon marriage, young couples build their own huts and live in them. Marriage ushers youths into the life of adulthood. Marriages are desired and long-lasting.

The family consists of the conjugal pair of husband and wife and their dependent children. The spouses do not live with the parents of either the husband or wife. Nevertheless, if the parents of either spouse are members of the same band, the couple is obligated to share their meat with their parents. Each conjugal pair forms a team within which domestic exchange takes place. The household is essentially nuclear; occasionally, however, relatives of either spouse may also be a part of the household. Families generally consist of monogamous unions, although polygyny is not completely ruled out. A marriage of a man to more than one wife has been recorded in Pygmy society,[17] and indeed a man with multiple wives may be accorded more prestige.

But such unions are more the exception than the rule. Marriage and the nuclear family structure of the Pygmies suit their hunting mode of existence by catering to the mobility requirements. Although spouses may change partners during their lifetime, couples are expected to settle down permanently after several short unions.[18] Procreation is the basis of marriage among the Pygmies. The terms "brother" and "sister" are extended to include any individual of a particular age group within the band. This fits in with marriage by exchange that prevails in Pygmy society. When a person wants to marry, he must find a mate for his "sibling" from another band.

To the Pygmy men, women are critical for survival, since women are essential partners in the hunting economy. Without a woman, a man is helpless. He cannot hunt, have a hearth or a hut. Therefore, the Pygmy men of one band demand that loss of a woman be compensated through replacement by a woman of marriageable age from another band. Thus, "sisters" and "brothers" are exchanged in marriage, and a band includes men and women linked together by a complicated network of kinship exchange. This process connects individuals into a reciprocal exchange, an element that has survival value for the simple society of the Pygmies.

Economic and Political Systems

The economic and political systems of the Pygmies are closely intertwined with their hunting and foraging method of existence. The impact of their technology is clearly discernible here. The Pygmies engage in both individual and collective hunting. However, even when an animal is hunted alone, the catch is not consumed solely by the hunter or his family. The Pygmies have reciprocal obligations within a network of kin and band members. Their economy can be described as one based on "generalized reciprocity," or shared reciprocity. This involves sharing of goods among people according to their needs without expecting anything in return. Goods are not exchanged at the time of economic transaction and the recipient is not obligated to return the favor to the donor immediately. This system of reciprocity ensures that everyone is provided for and that no one is without essential goods for survival. Indeed, very little value is placed on material possessions, since the resources essential for survival, including the tools needed to utilize the resources, are available to all. This factor has a leveling influence that eliminates distinctions based on rank or hierarchy. Consequently, all members are considered equal. The band leader or the experienced hunter is viewed only as the first among equals. Therefore, egalitarianism is built into an economy based upon generalized reciprocity.

All social systems entail some kind of economic exchange. Economic exchange is fundamental to human existence. However, the kind of exchange varies with the type and complexity of a society. Among the Pygmies, as among hunters elsewhere, meat is valued, and much of the economic activity revolves around obtaining and distributing this valued commodity. The sharing and exchange of meat brings individuals together in a network of domestic relations with other band members, as well as with Negro tribes living in the villages bordering the forests. Exchange of meat helps the Pygmies obtain such food as plantains and bananas, absent in the forest, from the villagers, as well as such vital items as metal spears, which are indispensable tools for hunting large animals such as elephants. Balanced reciprocity results, in a way similar to economic activity existing in industrialized societies where something is given for something that is received.

Within the domestic unit, economic activities of production and distribution of food require the conjugal pair to cooperate with other conjugal families. Although there is a division of labor between the sexes, both men and women engage in the hunting process. Women cooperate with men in the production of nets, an essential tool in collective hunting. Collective hunting involves the participation of both sexes. However, the critical aspects of net hunting are managed by women, and any wrong move on their part can spoil the hunt.[19]

The animals, once caught in the nets, are speared and killed by men. However, as we have established previously, men who own the nets and spear the animals caught in them do not have exclusive rights to the catch. Everyone involved in the hunting process shares the meat. The norm toward generosity ensures that the successful hunter will share the meat with the network of kin in residence in the band. Once the meat is distributed to individual households, each family is responsible for how it is consumed. The women, like their counterparts around the world, play a dominant role in the preparation and processing of food.

There is little differentiation of power and a minimum of hierarchy within the band. Positions of power and prestige are not limited through ascription, that is, they are not based on birth. Rather, achievement determines influence and prestige. Hunting skills are highly valued, and a successful hunter is respected and accorded high esteem. The qualities that enable individuals to assume leadership roles are not monopolized by any individual or group. Successful hunting skills enable individuals to achieve positions of influence in the Pygmy society; these positions of influence are not passed on nor are they restricted to a limited few. All capable persons can actually exert influence once they have demonstrated their skills as successful hunters. Furthermore, the hunting instruments or tools, the means of production in Pygmy society, are available in nature (liana and kusa vines, twigs and branches) or obtainable from the villagers (spears). Tools are available to all and therefore possession of them does not confer special privileges or power.

This technology is such that hunting skills are important, and this leads to a democratization that protects against the concentration of power by a limited few. The adults in the society who are actually the providers of the essentials of living, such as food, clothing and shelter, are "put in their place" and prevented by the youth and elders from dominating the society. This is done by playing down the contributions of these adults while underscoring the negative consequences of their actions, such as the "noise" they create during successful hunts and the jealousies and conflicts that occasionally mark adult relationships.[20] "Noise" refers not only to lack of silence, but also to conflict that can be highly disruptive of integrative bonds in a hunting band.

Pygmies are egalitarian, politically, economically and socially. The lack of inequality in their society is due to their egalitarian economy based upon reciprocal exchange. Experience and age, however, are given their due respect. Indeed, the importance of age is seen in situations of conflict and tension, where the elders are called upon to restore order and arbitrate between the grieving parties.

Religion and Values

The Pygmies do not have a complex system of religious beliefs. They worship nature in the form of the forest. Their survival depends upon it. They express their gratitude by singing to the forest, praising it for the food it provides in the form of plants, berries and animals. The forest is viewed as a benevolent friend, a protector that shields them from the outside world and showers them with kindness. The "molimo," the sacred object of the Pygmies, explains their religious life. A trumpet used to sing to the forest, it is treated with the same care, caution and devotion that is generally shown to supernatural beings.

The molimo represents the spirit of the forest and is invoked during times of major crisis to ensure that the forest stay alive and awake and protect its people from danger and harm. Through the molimo, the Pygmies communicate with the forest, praising or appeasing it according to the dictates of the situation, so that the forest can continue to provide comfort and the life they cherish. Their respect and deference to the forest are comparable to naturism, a form of religion in which objects of nature are worshiped. Some anthropologists have even suggested that worship of nature is the earliest form of religion.[21]

The Mbuti Pygmies live in the forest, love and protect it and, in turn, are themselves protected. Their relationship with the forest is one of peace and harmony. The forest is sacred and is used to sustain their minds and bodies. Any attempt to abuse the forest is resisted. This is noticeable in such things as their value system, their myth of origin and their relationship with the surrounding tribal villagers.

The Mbuti hunt animals only for food sufficient for sustenance. The contact with the tribal villagers has exposed them to new technology that they could use to increase their daily catch of game. Yet the Pygmies have only selectively accepted tools, such as the metal spear used to hunt elephants. The use of guns, which would have increased their hunting power tremendously, has been consciously rejected. Any attempt to exploit the forest for personal or commercial gains on the part of individuals is deliberately thwarted. During colonial times, the forest was kept off limits to outsiders such as the African villagers and the Belgian colonial administrators. The Mbuti have kept alive the tribal fear of the forest, which the latter regard as evil, a place for malevolent spirits. They have done this by telling the villagers stories of grotesque monsters that live in the forest.[22] It is through this technique that they have prevented the villagers from entering the forest for economic exploitation. The Mbuti even provided the villagers with needed forest food and game so that the villagers did not have to enter the forest themselves. In this way, the Pygmies have maintained its integrity. This tactic also prevented an overkill of the forest, for only the minimum required for subsistence was killed. In colonial times, the Mbuti also engaged in deliberate deception, giving incorrect information about the availability of the game to the Belgians in order to preserve the sanctity of the forest.

The myth explaining the origin of the Pygmies also upholds the sacredness of the forest. According to this myth, in the beginning people were immortal.

They lived on plants and vegetation abundant in the forest. It was killing that led to their mortality.[23] Hunting, while essential to their economy, is nevertheless viewed as a sacrilege, for it violates the integrity, purity and sanctity of the forest. It may be noted that this myth has striking parallels to the Christian doctrine of creationism.

Conclusion

The Mbuti Pygmies are the original inhabitants of the Ituri rain forest. The physical setting in which they live and the technology they possess are supportive of hunting and foraging. Since their technology lacks sophistication, they cannot harness or control nature; rather, they have to accommodate it for their food and sustenance. With very simple tools, they are able to exist on food naturally available in the forest. Their economy is simple and is based on a generalized exchange. Exchange of goods and services is not based on immediate reciprocity. Rather, one who can give does

The road through the forest on page 25 has been widened in the postcolonial period
Courtesy: Holt, Rinehart & Winston

Road running through the forest in colonial times
Courtesy: Holt, Rinehart & Winston

so, with the understanding that it will be reciprocated at a later time, when the present provider is in need. Upon this economy is superimposed an egalitarian political system. The hunting instruments (the tools of production) are accessible to all. Although there is some division of labor, both sexes participate in the hunting process. Positions of prestige and power are not limited to a few, but are achieved on the basis of hunting skills, and all people who demonstrate these skills wield influence. The forest, the source of their livelihood, is respected and worshiped for providing food and sustenance. Pygmy life is one of harmony with nature. The harmony and equilibrium noticeable among the Pygmies are due to the balance they maintain with their physical environment. Nonexploitative technology is certainly crucial to this relationship.

As hunters, the Pygmies are politically acephalous. Rulers and leaders are absent in their society. Although the Pygmies engage in economic and ceremonial relationships with the tribal outsiders, these relationships are mutually exclusive and dependent; each group obtains from the other what it cannot acquire in its own habitat, yet keeps its distance.

The Pygmies are now undergoing a transition. Consequently, their culture and way of life are changing. Nomadism is affected, and some Pygmies are adopting a way of life that ten thousand years ago began to revolutionize human existence. Pygmies are now lured by the temptation of material wealth available from the variety of peoples and industries emerging along the roadside. The roads leading to the forest are expanding to accommodate the increasing economic and commercial possibilities provided by the forest.

Change in the life of the Pygmies was first initiated indirectly by the Belgian colonial government when it required the villagers living around the forest to work on certain government projects such as road construction and to grow certain specific crops such as cotton.[24] These activities created a demand for labor as the able-bodied villagers were inducted into government road-construction projects. The pressure brought on the villagers to supply increased labor for the plantations introduced a major change in the relationship between the Pygmies and their African neighbors.[25] The increased demand for labor could only be met if the Pygmies helped work on the plantations. Thus, Mbuti were drawn into the plantation economy, which required them to be away from the forest for long stretches of time. They were unaccustomed physically and culturally to this kind of work. The intense heat of the sun did not suit them and they frequently became sick.

The colonial economy drew the Pygmies and the villagers into a new relationship. Although the Pygmies were left alone by the colonials, change began to infiltrate their lives and it accelerated with the passage of time. The harmony that had evolved between the Pygmies, the forest and their neighbors was disturbed. This was particularly evident in the postcolonial era, after the departure of the Belgian rulers. The country was torn by political factionalism as well as the demands for economic development. These factors penetrated the forest, which had hitherto maintained its integrity, with far-reaching consequences for its Pygmy inhabitants.

The colonial policy of exploiting the forest for economic and commercial purposes is perpetrated by the postcolonial government of independent DRC as well. The boundaries of the forest are receding as it is being cut down. Consequently, new forms of economic activities are emerging along roadside settlements. Furthermore, the cutting of the forest is also having a serious ecological and environmental impact. It has been noted that as the trees on the edges of the forest are cut to make room for road improvements, the shelter that the trees provided from the blistering hot equatorial sun is no longer available. As exposed road surfaces become hot and dry, they create an upward draft of hot air that drives rain clouds away.[26] This has begun to have a climatological impact, absent as recently as thirty years ago. It is likely that such environmental changes will affect the availability of the vegetation and game and thereby affect the people subsisting on them.[27] The indications are that economic changes have produced a changing flora and fauna which have made some forms of hunting unproductive.

The emergence of roadside settlements of villages, lumber mills, gold mines and administrative posts has done more to encourage the Pygmies to depart from their traditional way of life than the inducements of the villagers or colonial administrators. The Pygmies are gradually adapting to a surplus market-oriented economy. The increase in the number and kind of roadside communities has also increased the demand for forest items, particularly meat. Meat has become an important item of exchange. In the past, excess meat was traded for items available in the villages that the Pygmies did not produce themselves. Now the meat is sold for cash to middlemen who transport it to distant markets and earn ten times more than what they pay the Pygmies, who kill the game.[28] This demand for meat, however, is unmistakenly reducing the former autonomy of the Pygmies, who until recently killed only for subsistence and no more. Now surplus in meat is actively sought for economic motives. This incentive to kill for cash indicates a change from the former value system that guarded against overkill to protect the sanctity of the forest. It also marks the Pygmies' entry into a world of competition and hierarchy. This is unavoidable. As Pygmies kill animals for sale, they compete with each other. Access to cash gives them the opportunity to enjoy certain things unattainable to those without it. Herein is the basis of inequality absent previously. The pursuit of cash and material wealth also introduces Pygmies into a world of servitude.

Many times it is only by selling their services and labor that they can earn the cash needed to buy luxury items available in the markets sprawling around the roadside settlements. Thus, as the Pygmies succumb to the lure of cash and its benefits, they also automatically get drawn into the worlds of hierarchy, scarcity and inequality.

The Pygmies coexisted with other groups in the same environment for thousands of years because different resources were exploited by each group and shared with the other. Mutual reciprocity existed, allowing for a harmonious relationship. However, the recent immigration of new permanent settlers on the borders of the Ituri is threatening to upset the traditional relationship of the Pygmies to their ecosystem both within and outside the forest.[29] The new immigrants use the forest skills of the Pygmies for their own ends. For instance, they use the Pygmy knowledge of the forest to locate trees for needed lumber, much to the detriment of the forest. The long-term welfare of the Pygmies does not concern them. This is also noticeable in the market orientation of the newcomers. Their reliance on cash for economic transactions rather than upon reciprocal relations with the Pygmies places the latter at a disadvantage. Lacking cash, the Pygmies resort to menial and low-paying jobs in order to acquire it.[30] This places them in subordinate positions vis-a-vis the new immigrants. The continuation of this process could indeed usher in enslavement for the Pygmies and end the independence they previously enjoyed in the forest.

The DRC government's proclamation for emancipation of the Pygmies, granting them full citizenship and requiring them to abandon the forest, may lead to the final eclipse of the Pygmies and their way of life. The governmental directive that moved the Pygmies into permanent settlements outside the forest for purposes of taxes and census taking has been disastrous. As the Pygmies settled into model villages especially constructed for them, they experienced a variety of problems ranging from severe health problems to social abuse and deteriorating intergroup relations with the Negro villagers. The program had to be abandoned, since the Pygmies were not able to cope with the changes that this new style of living required. They returned to the forest.

The attempt to settle Pygmies into permanent villages may require careful social engineering. The Pygmies are now at a crossroads. To reverse the direction of change is impossible. However, as they make a giant leap into the twentieth century, crossing over cultural steps that humankind took several thousand years to make, we are provided a unique opportunity to witness a crucial chapter in the drama of technological forces and social change as it is actually happening.

Notes

1. Kevin Duffy, *Children of the Forest*, New York: Dodd, Mead & Company, 1984, p. 18.
2. Duffy, p. viii.
3. Colin M. Turnbull, *The Forest People*, New York: Simon & Schuster, 1961, p. 18.
4. Irwin Kaplan, ed., *Zaire: A Country Study*, Washington, D.C.: Foreign Affairs Studies of the American University, 1979, p. 12.
5. A. F. Putnam, *Madami My Eight Years of Adventure With the Congo Pygmies*, Englewood Cliffs, N.J.: Prentice-Hall, 1954, p. 32.
6. Marvin Harris, *Cannibals and Kings*, New York: Vintage Books, 1978, pp. 30-34.
7. Colin M. Turnbull, *The Mbuti Pygmies: Change and Adaption*, New York: Holt, Rinehart & Winston, 1983, p. 28.
8. *The Evolution of Political Society*, New York: Random House, 1967, p. 53.
9. Jean-Pierre Hallet, *Pygmy Kitabu*, New York: Random House, 1973.
10. ts, New York: Natural History Press, 1965, p. 22.
11. Turnbull, *Wayward Servants*, p. 23.
12. Turnbull, *Wayward Servants*, p. 97.
13. Richard E. Leakey & Roger Lewin, *People of the Lake*, New York: Avon Books, 1978, pp. 94-95.
14. Leakey & Lewin, pp. 94-95.
15. Richard B. Lee, *The !Kung San: Men, Women and Work in a Foraging Society*, New York: Cambridge University Press, 1979, pp. 318-319.
16. Lee, pp. 318-319.
17. Turnbull, *The Forest People*, p. 32.
18. Ernestine Freidl, *Women and Men*, New York: Holt, Rinehart & Winston, 1975, p. 23.
19. Duffy, p. 45.
20. Turnbull, *The Mbuti Pygmies*, p. 49.
21. D. N. Majumdar and T. N. Madan, *An Introduction to Social Anthropology*, Bombay: Asia Publishing House, 1963, p. 156.
22. Turnbull, *The Mbuti Pygmies*, p. 31.
23. Turnbull, *The Mbuti Pygmies*, p. 18.
24. Turnbull, *Wayward Servants*, p. 39.
25. Turnbull, *The Mbuti Pygmies*, p. 60.
26. Turnbull, *The Mbuti Pygmies*, p. 100.
27. Turnbull, *The Mbuti Pygmies*, p. 137.
28. Duffy, p. 80.

29. John A. Hart & Terese B. Hart, "The Mbuti of Zaire: Political Change and the Opening of the Ituri Forest," *Cultural Survival Quarterly*, Vol. 8, No. 3, Fall 1984, p. 18.

30. Hart & Hart, p. 20.

CHAPTER 3

Hunters of the Arctic: Eskimos

In the evolution of humankind, no single group of people has been as distinctive as the Eskimos.[1] They are hunters par excellence. The terrain that they inhabit is rough, unfriendly and dangerous; yet the skill with which they have adapted to and eked out a living from it is truly remarkable. The snow-laden, bitter and cold winter is their dominant hunting season, and the frozen sea, with its treacherous sheet of ice, is familiar ground upon which Eskimos have trodden year after year, having learned to distinguish from very early childhood the safe from the unsafe terrain. The hostile environment and the tools the Eskimos use to establish a subsistence from it influence their lifestyle, values and social organization. All of these in recent years have been affected by the onslaught of modern technologies and the economic forces accompanying them. The traditional Eskimo way of life is disappearing, leaving only vestiges of it as reminders of the way of life that dominated the arctic region for thousands of years. In the following pages, the story of the hunters of Northern Alaska is recounted and the changes brought about through the introduction of new technologies to their way of life are described.

The Modern Eskimos

The word Eskimo is a native Indian name that means "eater of raw flesh." It is a term that refers to groups of people inhabiting the arctic region, spread over territories bordering four nation states: Canada, Greenland, the Soviet Union, and the United States.

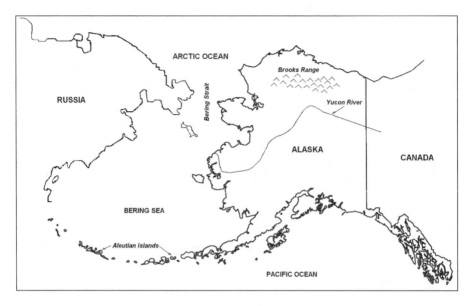

Map of Alaska

There are Eskimos in Greenland, Northern Canada, Northern, Western, and Southern Alaska and the tip of Siberia. While these groups of people cover a wide range of territory, an examination of their culture reveals that there are striking similarities in all of their customs and languages.[2] There are different tribes of Eskimos, and the names by which they are identified vary from region to region.[3] The people of Northern Alaska are known as Inupiat and they speak the language known as Inupiaq, a language also spoken by Eskimos of Greenland and Canada. Our study here is concerned mainly with the Inupiat people of Northern Alaska, and since there is no word other than Eskimo that collectively refers to the native people of Northern Alaska, it is used in the discussion that follows.

There are forty thousand Eskimos in Alaska[4] and they represent less than half of the total population of Eskimos distributed through the territories of different nations. Of this, only a small portion are North Alaskan Eskimos. Although they move to temporary locations when hunting or fishing so demands, most North Alaskan Eskimos live in fairly large, year-round coastal villages.[5] The abundance of sea and land mammals together with other forms of marine life and birds has turned Alaska into a rich hunting territory that Eskimos of the region have exploited only with tools designed from the bones and skins of the animals killed. With quite simple and primitive tools, the Eskimos of Northern Alaska have been able to procure their year-long requirements of food until this balance with their environment was disturbed through contact with Western European civilization.

The Physical Setting

The Arctic Coast of Northern Alaska is a long stretch of coastline extending from the Barter Islands in the east to Point Hope in the west. The northernmost point of this coastline is Point Barrow, which lies midway between the two extreme eastern and western points. The eastern section is geographically quite dissimilar from the western part, giving rise to different hunting conditions and population distribution. In fact, the eastern part of the arctic coastline is thinly populated, while the part west of Point Barrow is densely populated.[6] A striking feature of Northern Alaska is its chains of mountains. The most northerly, the Brooks Range, forms a continental divide separating the treeless flat tundra of the far north from the thick forests of the central Alaskan plateau. The Brooks Range is the highest mountain range within the Arctic Circle and stretches six hundred miles from the Canadian borders to the Chukchi Sea. This range divides the waters flowing northward into the Arctic Ocean from those flowing southward, which are eventually emptied in the Bering Sea. These mountains have several well-disguised passes through which caribou make their annual migrations.[7]

The Alaskan coastal plains consist of typical arctic tundra. In the summer, streams and rivers are seen flowing before emptying into the sea. The summers are brief, lasting from six to twelve weeks. At this time, the temperatures rise above freezing and daylight lasts for up to twenty-two hours. Throughout the summer, flowers, mosses, grasses and lichen spring up, turning the tundra into a rich hue of green. The coastal plains also attract a variety of wildlife, such as foxes, rabbits, herds of caribou and migrating birds. Seals and walrus bask in the sun. In late September, the sea begins to ice over and the long winter freeze begins. From October to July the waters are locked in ice and the land lies frozen. At this time, the arctic animals, except for the seals and walrus, migrate southward or hibernate. Temperatures may fall as low as -30 to -50 Fahrenheit.

The Alaskan winter provides an environment of rich marine resources that the Eskimos have exploited. In fact, the arctic coastline can be separated into two divisions, the land and sea; both offer a rich and abundant supply of animals. In the sea, the movement of snow and ice triggered by winds or currents creates cracks and openings of water; here, mammals can swim to the

surface for a breath of air. These movements of ice and snow provide opportunities for hunting, but also create dangers which, if unheeded, can lead to
risk of life. Therefore, not only must the Eskimo observe the behavioral
characteristics and mobility patterns of the game he pursues, but he needs to
know the behavior of the ice-covered sea that supports the rich forms of life
that sustains him and his people.[8] The stages of ice development as the winter
approaches progress and recede; their different kinds of snow, strength and
thickness, and wind patterns are all keenly observed by the Eskimos. Their
success as hunters depends on mastery of this body of knowledge. Over the
centuries, Eskimos have acquired this knowledge and developed tools and
skills to ascertain the conditions under which they can safely pursue the game
upon which their life depends. It is this ingenuity that has permitted them to
be successful hunters in an environment that is generally considered to be
uninhabitable. It is thus not surprising that there are one huundred words in
the Eskimo language that distinguish the various kinds of snow that shroud
the arctic surface during much of the winter and spring.

The Hunting Economy

Arctic sea and land animals have been important to the hunting economy of
the Eskimos. For centuries, the Eskimos have hunted such prized animals as
the bowhead whale and, the polar bear and lesser animals such as the beluga,
caribou, wolf, walrus, fish and land and water fowl. These animals have
provided the basis of their hunting and gathering economy. However, there
have been differences in the subsistence patterns of Eskimos according to the
animals hunted.[9] Eskimo hunting patterns have been determined by the
availability of the animals and their migration habits. A look at the aboriginal
North Alaskan calendar illustrates the importance of the different animals to
their economic and social life. The Eskimo year is divided into two main
seasons, winter and summer, and each is then further divided into a number
of subseasons.[10] During winter, seal hunting is the main hunting activity, as
other game is scarce at this time. January and February are the coldest winter
months and during this time seals are hunted in their breathing holes, made
by the seals when the ice is just forming. Ice-edge sealing also takes place at
this time. In fact, it is at this time that the most intensive hunting for seals
along the edges of open leads, that is, channels of water in fields of ice, is
carried on.[11]

An Umiak used for whale hunting
Artist: Barbara Boucher

Breathing-hole seal hunting requires a long, patient wait in the cold winter and is not as productive as ice-edge sealing. At most one or two seals can be killed at one time with harpoons in the breathing holes. Only old men still engage in breathing-hole seal hunting.[12] Since rifles have been adopted, harpoons are no longer used. The introduction of rifles brought about changes not only in the human-animal relationship but also in the hunting activity itself. Use of rifles increased the number of seals killed at one time and made ice-edge sealing a more important aspect of hunting than it had been formerly.

Springtime ushers in the whaling season. It is at this time that whales are seen migrating northward. They pass through narrow channels in the ice close to the shore, moving along different points of the coastline from the Bering Strait to the Beaufort Sea. Both the majestic, bowhead whale, measuring up to sixty feet long and weighing up to seventy tons, and the smaller beluga or the white whales are hunted in springtime.

Traditionally, whales were hunted with harpoons and lances. Harpoons are hunting weapons whose primary use is to wound and then attach the animal to a line, at the other end of which is the hunter himself.[13] Skin-covered boats known as umiaks are used in pursuit of whales. Whale hunting was and still is communal in nature. An umiak is usually twenty to thirty feet in length, and can accommodate a crew of up to eight people and their hunting weapons. A crew includes the captain or steerer, a shoulder-gun man, a harpooner or striker and three or more paddlers. As soon as the whales are spotted, umiaks are used to chase the whales through the water; the crew paddles silently but softly in order to come close to the whale's left side, allowing the harpooner to pierce its vital spot with his right hand. A sealskin float attached to the harpoon line is tossed by a man seated just behind the harpooner when the harpoon is thrown. If the whale comes close to the boat, two men can kill it.[14] Other crew members, of course, are needed to row the boat and steer it.

Traditionally, single-manned boats known as kayaks were also used in chasing whales and hunting walrus. A solitary hunter could quietly slip into the water in pursuit of whales or walrus, but hunters were often endangered when whales toppled the boats. Nevertheless, kayak hunting prevailed in Eskimo territory in the past. Umiak hunting, on the other hand, still continues.

Whales are also hunted from the ice edge. In the spring, when the whales make their annual northward migration, the Eskimos wait in their whaling camps to listen for their blowing. Upon hearing the whistling sound of water blowing, the hunters station themselves close to the water's edge, behind the high piles of ice so as not to be seen, and ready themselves to shoot at the first opportunity.

Whale hunting depends upon the condition of the ice and weather. It demands great investment in time, resources and labor, as it involves hard work and a long wait for the crew on cold, bitter nights in the hope that a whale will be killed. The successful crew gets rewarded with plenty of meat for all; prestige is conferred on the hunter or hunters responsible for capturing and killing the whale.

Whaling camps are set up along the shores of the open lead, where the crew members rest, eat and watch for the whale. The camps consist of tents, with snow and ice blocks used as anchors for the side ropes. The tents measure about ten square feet and are six feet high. These camp sites are used primarily as whale-watching posts and have to be moved if the condition of the ice and snow warrants it. A flat area is essential near the campsite; here, the dogs used for transportation are lodged and the game is butchered and divided.

Whale hunting involves a cooperative effort between the members of the hunting team. There is no other way that a huge whale can be killed and retrieved. There are rules that dictate the sharing process once the marine mammal has been killed and brought back on shore. The custom governing division of a whale is based on order of participation in the kill. The boat crew that successfully kills the whale receives the choice parts. Next in order of privilege comes the other boat crews who have helped in the kill. Finally, the remaining shares of meat go to those who first touch the whale after it is killed.[15]

The captain of the boat is responsible for the clothing and equipment of his crew. In Eskimo society, all whalers are respected, but a special kind of status is granted the boat captain of a whaling crew. He is a provider of abundant meat and, as such, exacts enormous respect and gratitude from his people. It is he who is responsible for overseeing the details of final prepara- tion prior to the whale hunt. Futhermore, the captain is required to be generous and gains honor from his generosity. In fact, custom dictates that he receive no share from the first whale he kills, although he is required after his first kill to give gifts to those relatives who have a right to ask for them.[16] His primary reward remains the enormous prestige he receives as the successful whaling captain. On later hunts, however, he receives the first priority of choice. Traditionally, the crew consisted of extended family members, though this is not as common now as it was in the past.[17] Now the captain of the boat recruits his crew from the open market, tempting the skilled hunters to defect from the group to which they traditionally belong, while ensuring that skilled hunters of his team do not forsake him to join other hunting boat crews.[18] Monetary and other rewards are used as incentives to retain skilled hunters as crew of a whaling team.

Whale hunting is an important economic activity and many Eskimos still pursue it. However, today the number of Eskimos with permanent salaried jobs in government and private agencies is increasing, bringing about a change in the integrative bonds that once linked these people together in their cooperative struggle for survival in the harsh arctic environment.

After the spring whaling season, the Eskimos turn to hunting of walrus and seals. Walrus hunting really begins in July after the land-fast ice begins to break and the pack becomes somewhat scattered. Walrus are hunted by boat crews in much the same way as whales,[19] although Eskimos do not hunt them in open waters because they are dangerous.[20] Since walrus hunting is considered as dangerous as whale hunting, both are preceded by a prayer. A Christian prayer asking for protection and a successful hunt is recited by one member of the hunting crew, while others listen in silence.[21] After the prayer is over, these formidable beasts, sighted riding on ice bars, are hunted along the shores. Once the hunter is within shooting range, the walrus are shot and harpooned for retrieval. Several men are required to haul the animal onto land, where it is butchered. The number of animals killed is limited by the capacity of the boat to transport them. Once the killed walrus are hauled onto the solid ice, men do the butchering. It is tedious work and exclusively done by men working in pairs. Traditionally, walrus were hunted with harpooons and lances. Now rifles are used to kill them; harpoons are used only when the beast is completely immobilized. Harpoons ensure that the game is not lost in the water, as it secures to the animal.

Caribou is the most significant land animal hunted in the summer and fall. During these seasons, herds of caribou are on their annual migration northward to the Arctic Circle. With the coming of the fall, masses of herds move southward toward the mountains and forest, leaving a fraction of their kind to be stalked by men. The Eskimos hunt the caribou by remaining concealed close to the migratory paths; sometimes traps and snares are also used. Often hunters follow the tracks of caribou, approaching them with caution so as not to forewarn the animals of their presence. The hunters control their team of dogs, leaving them at a distance, proceed on foot within a comfortable shooting range and then take their aim. As many animals are killed as can be hauled onto the sled and carried back home. The animals are first skinned and dismembered at the joints; the flesh is removed along with the other delectable parts of the freshly skinned animals. The animals are then loaded on to the sled. Often, following the traditions of native Indians, caribou that cannot be transported are cached at the hunting site and taken back at a later time.[22]

Traditionally, caribou were hunted with bows and arrows. With the introduction of rifles, these traditional weapons have been replaced. Rifles make hunting easier and have increased the number of animals that can be killed each time. But this technology has also affected the social philosophy of hunting by reducing the need of interdependence between hunters and diminishing the prestige of the hunter.[23]

Birds, such as ducks, ptarmigan, and gulls, are also hunted. Some of these are caught by women and old men in baited snares. In addition, fish is also caught when the season allows. Of course, these sources of food are not afforded the same kind of importance in the economy as big game hunting on land and sea.

Hunting is life and life is hunting for the Eskimos of Northern Alaska, for upon it depends their survival as a group. Eskimos never take food for granted. Given the struggle for existence between man and the animals that abounds in the region, the Eskimos have devised ingenious reasons whereby possessions that are superfluous are considered undesirable. Functionality of possessions is important and governs their social existence. All items of culture are geared to a purpose. Such attitudes have made them successful in spite of the harsh and uncooperative environment that surrounds them. The significance of function is evident in the social organization and norms that govern their everyday life and relations between people. It is to these that we turn now.

Social Organization

There are two different groups of Eskimos in Northern Alaska:[24] the inland Eskimos who depend principally upon caribou hunting for their livelihood, and the maritime hunters whose chief source of livelihood derives from sea mammals such as whales, walrus, and seals.

These groups engage in trading with one another so that each provides the other with necessities not available in their own habitat. Moreover, the social stucture of both groups is fairly uniform.[25] This uniformity can be attributed to the communal nature of the hunt wherein the demand for adequate food supply necessitates that the members of each region engage in cooperative efforts if the large game upon which their livelihood depends is to be successfully hunted. Thus, in both groups one notices that social mechanisms are at work that bring individuals together at hunting sites. It is here that extrafamilial social patterns arise.[26]

The basic unit of social organization is the nuclear family, consisting of the mother, father and the children. It is, however, not an isolated unit. A household consists of several nuclear families whose members are related through affinal or blood ties. In practice, most nuclear families are part of a larger residential group, a household of extended family members. Aboriginally, however, the bilateral, extended family was the basic unit of social structure among the North Alaskan Eskimos.[27]

Within the household, the parents have the primary responsibility of preparing their children for adult roles. Other adult members are nevertheless important. Kinship is important in organizing and influencing social relationships. However, practical considerations such as propinquity, residence, frequency of social interaction and mutuality of interest are more important in maintaining kinship ties than the mere recognition of the relationship.[28] Since the hunting economy depends upon constant cooperation between kin and quasikin, it fosters economic interdependence between them. In fact, economic interdependence and mutual cooperation go hand in hand and are mutually reinforcing. Cooperation and hard work determine the success of a household; therefore, its members are called upon to engage in mutual aid and reciprocal obligations. These prevail only so long as interdependence is the key to survival and success. Opportunities allowing an individual to be independent of the cooperative kin network lessen the need for economic interdependence. These have been provided by the introduction of rifles and opportunities for individual wage labor.[29]

Within the family, the husband and wife form an economic team. The hunter, as the economic mainstay of the family, is a pivotal figure within the family, as well as in the larger society, especially if he is known as a successful hunter. In principle, and in practice, hunting of big game such as sea mammals and caribou is predominantly men's work. Indeed, traditionally, women could not even accompany a hunting crew on its whaling expedition.[30] Women did participate in communal caribou hunting, where their task was mainly confined to driving the caribou in the direction where the snares were laid and the hunter waited in ambush.[31] They also sometimes helped their husbands in sealing, but only in those exceptional circumstances when men were unavailable or reluctant to hunt. Reluctant men are subject to ridicule and derided for not assuming their appropriate male roles. In such situations, these men are assigned chores generally performed by women.[32]

The primacy of hunting in the economy of Eskimo life, and the predominance of men as hunters, have definite social consequence for the status of the sexes in the community. Since men have the active role of procuring food, clothing and fuel, essential for subsistence, male children are highly prized. Indeed, aboriginally the Eskimos practiced female infanticide. It was men who were burdened with the prime responsibility of obtaining food in an environment full of risk and danger. They had to acquire the knowledge and experience to become successful hunters. These skills provided an opportunity to gain positions of social prominence in the community and acquire respect and power. The captain of the whaling crew or caribou hunt is always a man. As such, it is he who distributes the meat from the hunt within the household and community. This act gains him the gratitude of all who benefit. Since they are nonhunters, women are excluded from this complex network of the acquisition and distribution of the desired prized meat.[33]

The differential positions and privileges of men and women are reflected in the norms and practices of marriage. Traditionally, marriages are arranged by parents, and a man who is honest and a good hunter is desirable as a prospective son-in-law. A husband and wife form an economic team and economic considerations are paramount in the settlement of a marriage. Even so, within the household and marriage, men make the decisions and women obey.[34] A woman is regarded as the sexual property of her husband so much so that her consent is not deemed necessary when he decides to share her sexual favors with other men.[35] Indeed, exchange of spouses is one method of creating economic alliances between men and helps foster cooperative patterns of interactions and support. Exchange of sexual partners draws men into dependable, long-term relationships for mutual aid. The men who have shared a woman sexually are brought into a partnership, a formalized reciprocal relationship with each other.[36] These extensive ties allow families and individuals to move about on long- and short-term visits and thus enable the Eskimos to adjust their group structure according to the availability of the resources and personal preferences of the individuals comprising the group. These bonds can be terminated just as marriage can. However, as long as these relationships are in effect, they are treated as kinship bonds and men can request assistance from one another when the situation so arises. This pattern of mutual aid is even extended to their children. Because spouse exchange usually occurs between communities, the main benefit of this relationship is the extension of hospitality during travel in a strange territory.[37]

Male dominance in relationships between sexes is further evidenced in premarital sexual encounters. Although general aggression, both physical and verbal, is frowned upon because it can lead to disharmony in the community, sexual aggression against women is tolerated. A man can force a woman into sexual intercourse. This does not jeopardize her prospects for marriage, even if the act results in her becoming pregnant. However, a promiscuous woman earns a bad reputation and consequently is not desirable as a wife.

As stated earlier, within marriage, the husband and wife form an economic team and though marriages are not devoid of mutual affection, economic cooperation remains an important aspect of the marriage relationship. A lazy woman is unlikely to be sought after in marriage, since in the arctic environment a man needs a wife to do the chores within the house so that both can survive. This economic cooperation is well illustrated in the division of labor that exists between spouses. A woman's prime responsibility is to maintain the household. Although she aids her husband outdoors in various hunting tasks, the largest part of her daily activities are directed toward food preparation, cooking, preparing and sewing skins and tending babies. A man spends most of his time hunting and making the tools and weapons required for it. It is only now, with the availability of wage employment, that romance and courtship have become important in the process of mate selection, and the economic skills of hunting, once considered indispensable in a mate, are no longer thought of as essential.

Beyond the family, the hunting groups and karigi, or men's ceremonial dance and clubhouses, are important social units within the social structure. Like marriage (which draws individuals and families in a relationship of economic cooperation), hunting groups also bring people, many of whom are unrelated, into economic partnerships. Success of a hunting team demands that skilled and experienced individuals come together to form a crew. Such persons are not always available in the circle of kin, so that a hunt leader sometimes has to go outside this circle to entice an individual of good hunting reputation to join his crew. Societal mechanisms are available for bringing this about. Wife exchange, noted earlier, is one means of doing this. Gift giving and bribery are others. In fact, harpooners use gifts as leverage before deciding to render their services to a crew captain.[38] The harpooner, whose skills often decide the success or failure of an entire whaling crew, is tremendously important to the operation. Thus, he cannot be taken for granted. A competent whaling captain must attract a skillful and experienced harpooner, and such societal mechanisms afford an element of certainty to the members of a hunting team, confident in the knowledge that they are drawn into a reliable and secure relationship for long periods of time. Enlisting skilled individuals into the hunting team depends on the successful recruiting of the captain. A test of his leadership qualities rests upon the fact that he has to act in such a way that even when he draws a member from another hunting crew, his action does not lead to any retaliatory behavior from the "losing" team.[39] The hunting group is bound together for a primary economic purpose in a society where economic achievement is essential for survival. Intergroup rivalry that interferes with economic goals is avoided or controlled.

The hunting group is not the only social unit beyond the family and circle of extended kin. The men's house, or *karigi*, is of great social significance to the life of the Eskimos. It includes several hunting crews, and an individual is a member by virtue of his membership in a given hunting crew whose *umiak* or captain has joined or founded the karigi. An individual identifies with the community in which the men's house to which he belongs is located. He maintains good relations with its entire membership. The house is a social meeting place for men. It also provides a center for observation of religious ceremonies associated with hunting activities. Karigi, however, lost their importance with the coming of Christian missionaries to Northern Alaska and they have now almost disappeared.[40]

Technology and Society: Changing Scenario

Traditionally, the Eskimos have depended primarily for their livelihood upon hunting. In an environment that is harsh and hostile, they have survived for thousands of years, having learned to maintain an equilibrium between themselves and the surrounding habitat. Their technology, simple but efficient, has provided challenges to individual hunters to demonstrate their hunting skills and thus validate their positions as prized members of their communities. The status of a successful hunter has to be achieved, which provides ample opportunities for a person to pursue avenues for individualism. Yet, cooperation between individuals has been the key to survival and hence has been valued. In fact, Eskimo society had arrived at such a harmonious balance between the individual and collective that while every individual was given the opportunity to develop and realize his potential as a skillful hunter, the ideal and reality of cooperation continued to permeate social institutions and values. Individual achievement was possible because of the cooperative endeavors that sustained it.

While hunting still continues, changes are nevertheless taking place. The introduction of the rifle and other steel implements has initiated changes in the social spheres of life. Although the rifle has made hunting easier, it nonetheless has reduced the need for sharing and cooperation among kin and at the same time has lessened the prestige of the hunter.[41] Previous to its introduction, when Eskimos hunted with bows and arrows, they engaged in corral caribou hunting during the spring and fall migrations of the herds. This necessitated the formation of bands to exploit the seasonal migration of caribou. Just before the spring or fall migrations, the hunters had to get together to plot their plans, to ensure that their equipment was available in sufficient numbers and in good order. They also assigned particular individuals tasks to assure the most effective use and deployment of men within the band.[42] Today, this is no longer the case. The rifle has made hunting an individual activity, and now every man hunts alone or in small groups of close kin and friends. The individual hunts solely for his own household.[43] The introduction of the rifle has stimulated other changes as well. For example, it has also eliminated the traditional importance of customary religious practices, since its efficiency obviates the necessity for supernatural intervention, thereby calling into question the efficacy of ancient rituals and taboos connected for millennia with successful hunting. Traditional methods of societal control consequently have also been affected.[44] One notable example is the declining role of shamans after white whalers consistently killed large numbers of whales even without observing the traditional taboos associated with whale hunting.

Today, whaling is still an important economic activity. Changes in whaling technologies have brought changes in partnership patterns.[45] The introduction of the dart gun has somewhat reduced the risk inherent in whale hunting by making the dangerous chase unnecessary. However, whale hunting still involves specialized cooperative endeavors and some of the traditions associated with the practices of crew recruitment and division of meat distribution persist. In fact, the force of tradition in whaling was tested when the Eskimos banded together and succeeded in fighting the effort of the United States Congress to ban the killing of bowhead whales.[46] Whaling is a vital part of the Eskimo culture and therefore continues. However, with modern whaling technologies, it is only the man with wealth who can afford to organize a boat crew. Outfitting a boat requires a cash outlay of seven hundred dollars[47] and this has limited the number of boat crews that engages in whaling.

As new economic opportunities have become available to the Eskimos with the discovery of energy resources and the establishment of military installations, they have come into contact with American culture. This, in turn, has brought a change to the traditional way of life of the Eskimo people. New material goods, recreational patterns and religious observances indicate the scope of the influence that experience and contact with American culture has brought. This has had mixed consequences. On the one hand, it has meant controlling rather than adapting to the environment, thereby making life easier. For instance, the tradiontional frozen sod igloo, with underground entrance tunnel and skin covered walls, has been replaced by homes with electricity.[48] But the material revolution has been accompanied by a social price—attenuation of communal life, resulting in boredom, alcoholism and family violence.[49] Participation in a wage economy has brought material affluence and an improved way of life that many young Eskimo aspire to. On the other hand, it has also meant movement away from traditional participation in such communal concepts as sharing and cooperation among kin groups. This has created difficulties, because the Eskimos are not used to a cash economy. These difficulties can be seen in the response of a wife whose husband cannot find a job, although she has one.[50] She notes that her situation can provoke an identity crisis in her spouse, who, although he can fill the freezer with meat, is no longer able to live up to the masculine ideal of the husband as the sole economic provider for his family's needs.[51] The new cash economy, with better job opportunities for women, has disrupted the traditional role patterns of work between the sexes, leaving men vulnerable and struggling.

The American influences have imposed upon Eskimos a cultural dilemma. Participation in the cash economy requires that children go to school to acquire the necessary skills for successful entrance to the job market. On the other hand, it also means an end to the Eskimo family's nomadic pursuit of the arctic animals upon which their traditional livelihood depended. A society that is egalitarian in nature is unlikely to remain so when prestige no longer flows from personal skills of the individual hunter but depends upon gaining positions valued in the market, the availability of which is limited. As the Eskimos have been drawn into a market economy, they have had to learn to accept market values and institutions. In so doing, they have moved from an egalitarian social organization to a stratified one. Consequently, Eskimos of Northern Alaska are drawn into a system of social hierarchy marked by inequities of position and privileges.

Eskimos are now at a cultural crossroad. There is the pull of the American market economy and all that it implies threatening to engulf their traditional millenia-old value system and way of life. Confronting its inroads is the resilience of the traditional culture, which attempts to resist this tide of Americanization. In this regard, Eskimo organizations have sprung up across the Arctic Circle, demanding cultural, economic and political autonomy for themselves. Only in this way, they maintain, can Eskimos accommodate to the opportunities that confrontation with a modern market system provides without being totally eclipsed by it.

Pygmies and Eskimos: Some Comparisons

For thousands of years Pygmies and Eskimos have adapted to their respective ecosystems through strategies of subsistence that allowed them to carve out a living from their respective environments in spite of rudimentary technologies. The technologies of both groups evolved from their individual environments and were suited to them. Recent contact with new technology and the value systems of the outside world has initiated changes in the indigenous worlds of both the Pygmies and the Eskimos. Consequently, new social patterns and values previously unknown to both societies have emerged. Social forms and values formerly believed to be indispensable have also eroded. The following brief comparison of the two societies outlines the similarities and differences in both cultures and their responses to the changes affecting them, as a result of the impact of Western technology.

Pygmies lived by hunting and gathering. Hunting was primary in the Eskimo economy. These economic realities affected the structure of the relationship between the sexes in both cultures. In Eskimo society, men were the providers, since hunting was principally a male activity. Women's contribution to the economy, although important, relegated them to a subordinate position. The Pygmy women, by contrast, were men's economic partners. Their skills were essential for successful hunting. Therefore, they had an egalitarian relationship with men. Despite this difference between these two societies, there is much that was common. For both the Pygmies and the Eskimos, cooperation was crucial for survival and success. It was necessary and stressed. Generosity was emphasized. Indeed, the pressure toward generosity was a mechanism assuring that resources such as food would be shared by all members of the group. Individuals and families were linked together in a chain of mutual reciprocity and interdependence. This interdependence was extended outward to include other groups that engaged in different pursuits. Thus, the Pygmies were locked into a relationship of mutual interdependence with the non-Pygmy villagers living around them, each supplying the other with what it could not obtain on its own. Similarly, the coastal Eskimos traded and exchanged with inland Eskimos goods and items that each needed from the other. Cooperation, sharing and generosity were important aspects of behavior. They were also important values. Accumulation of material possessions was discouraged. Requirements of subsistence dictated this. Possessions were burdensome in a society where agility and movement were important. This, of course, created egalitarian relationships between individuals and between groups. No one possessed anything extraordinary that set them apart and could give them an advantage over others.

The Pygmies and Eskimos are now facing new challenges and opportunities as they confront the forces of a market economy. The Pygmies, like the Eskimos, are at a cultural crossroad. They maintained their egalitarian social structure even when the surrounding villagers were invaded by the forces of Belgian colonialism. Unlike the Eskimos, the Pygmies rejected the use of the gun in order to preserve their way of life in the forest. However, with the nationalist government coming to power, the Pygmies have been drawn into a market economy with wage labor. As the lure of material wealth draws the Pygmies into an external market, they succumb to the temptation of surplus. In so doing, they also accept the principles of social hierarchy and inequality, elements that disrupt their egalitarian society. While change is unavoidable, there is still for the Pygmies, as with the Eskimos, a division between the young and old concerning the desirability of such changes.

Eskimos were exposed to change when the white man arrived on their frontiers for commercial exploitation. The introduction and acceptance of the gun by the Eskimos brought significant ecological and social changes in their lives. Cooperative efforts in some kinds of hunting gave way to individual enterprise. Sharing declined. More recently, exposure to a wage economy has further introduced the element of individualism in their lives. As people obtain regular employment, former patterns of social living become impractical. Consequently, cooperative ties between individuals and families have decreased as people have become less dependent on each other. A wage economy has resulted in cash income and its benefits. Increase in material wealth has brought about a social hierarchy and inequality among people. While making life predictable and comfortable it has also been accompanied by increases in social problems such as alcoholism.

Both the Eskimos and Pygmies are now drawn into a market economy and are likely to be affected by its fluctuations. To expect that they will preserve their culture and tradition in an unchanged form is unrealistic. The future will demonstrate the extent to which each society will successfully retain its identity while accommodating to the forces of a modern economy that requires the individual to compete rather than cooperate with others in the art of living.

Notes

1. Herbert Wally, *Eskimos*, London: Collins Publishers and Franklin Watts, Inc., 1976, p. 7.
2. N. M. Giffen, *The Roles of Men and Women in Eskimos Culture*, Chicago: University of Chicago Press, 1930, p. vii.
3. David Boeri, *People of the Ice Whale*, New York: E. P. Dutton, Inc., 1983, p. xi.
4. Wally, p. 17.
5. Norman A. Chance, *The Eskimo of Northern Alaska*, New York: Holt, Rinehart & Winston, 1966, p. 5.
6. Richard K. Nelson, *Hunters of the Northern Ice*, Chicago: The University of Chicago Press, 1975, p. 3.
7. Chance, p. 7.
8. Nelson, p. 3.
9. Dorothy J. Ray, *Ethnohistory in the Arctic: The Bering Strait Eskimo*, Ontario: Limestone Press, 1983, p. 175.
10. Nelson H. H. Graburn, *Eskimos Without Igloos*, Boston: Little Brown & Company, 1969, p. 35.
11. Nelson, p. 249.
12. Nelson, p. 233.
13. Wally, p. 90.
14. Nelson, p. 217.
15. Richard K. Nelson, *Shadow of the Hunter*, Chicago: University of Chicago Press, 1980, p. 83.
16. Nelson, *Shadow of the Hunter*, p. 85.
17. Chance, p. 38.
18. Wally, p. 96.
19. Chance, p. 40.
20. Nelson, *Hunters of the Northern Ice*, p. 354.
21. Nelson, Hunters of the Northern Ice, p. 362.
22. Graburn, p. 48.
23. Chance, p. 2.
24. Ernestine Friedl, Women and Men: An Anthropologist's View, New York: Holt, Rinehart & Winston, 1975, p. 39.

25. Robert F. Spencer, "The Social Composition of the North Alaskan Whaling Crew," in Alliance in Eskimo Society, Lee Guemple, ed., Seattle: University of Washington Press, p. 110.

26. Spencer, p. 111.

27. Chance, p. 48.

28. Spencer, p. 110.

29. Nicholas J. Gubser, The Nunamiut Eskimo Hunters of Caribou, New Haven, Conn.: Yale University Press, 1965, p. 77.

30. Lael Morgan, And the Land Provides: Alaskan Natives in a Year of Transition, New York: Anchor Press/Doubleday, 1974, p. 75.

31. Gubser, p. 174.

32. Giffen, p. 2.

33. Friedl, p. 41.

34. Friedl, p. 43.

35. Friedl, p. 44.

36. Spencer, p. 113.

37. Gubser, p. 65.

38. Spencer, p. 116.

39. Spencer, p. 118.

40. Chance, p. 54.

41. Chance, p. 2.

42. Gubser, p. 142.

43. Gubser, p. 142.

44. Chance, p. 2.

45. Spencer, p. 113.

46. Morgan, p. 96.

47. Morgan, p. 50.

48. Samuel Z. Klausner & Edward F. Foulks, Eskimo Capitalists, Oil, Politics and Alcohol, Totawa, N.J.: Allanheld, Osmun & Company, 1982, p. 43.

49. Priit J. Vesilind, "Hunters of the Lost Spirit," National Geographic, Vol. 163, No. 2, 1983, p. 157.

50. Vesilind, p. 157.

51. Vesilind, p. 157.

CHAPTER 4

Hoe Cultivation:
The African Example

Introduction

The invention of plant cultivation, that is, deliberate planting of roots or seeds for growing crops, occurred around ten thousand years ago. In this chapter we will discuss the system of plant cultivation known as horticulture, where the cultivation of the soil and planting are done by humans without the aid of plow drawn by draft animals or machines. Horticulture is a method of plant cultivation used in areas of tropical forests and in savanna regions.[1] Horticulturists are "tribal" people found in different parts of the world, including Africa, from which our case is drawn.

Many East African communities engage in horticulture. Kenya, the case study we will examine, is one of these communities. Its method of soil cultivation involves a very simple technology: the hoe and digging stick, which are used to prepare the soil for planting and cultivation. This method is also known as shifting cultivation. So common has this method of agriculture been in Africa, that it can easily be identified as a land of shifting cultivation, as opposed to Asia, where the main implement in agriculture has been the plow.[2] This is significant, as the use of digging stick involves a relationship between land and people quite distinct from what prevails in societies using plow cultivation.

The African agricultural system presents a dramatic case study for examining the human factors in technological change and the technological forces in human development. Since Kenya is illustrative of a shifting cultivation, we will discuss its society, which depended upon hoe technology for its primary existence, in the following pages. In so doing, we will delineate the interrela-tion-ships between technological, societal and value forces. The conditions

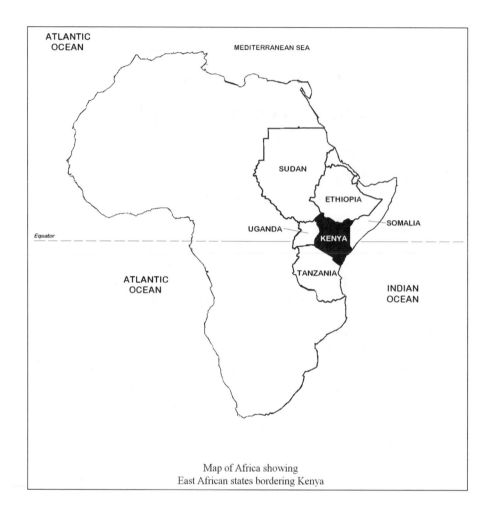

Map of Africa showing
East African states bordering Kenya

that prevailed prior to European contact and the changes introduced since
then will be discussed to emphasize the outcome resulting from the confrontation of emphasize the outcome resulting from the confrontation of technological, societal and ideological forces upon Kenyan societal development.

Physical Features

East Africa is a land of great diversity. Its variety is discernible in climate, population, economic and political conditions. It can conveniently be divided into five national regions that cut across political boundaries.[3] The coastal fringe has for centuries had contact with the outside world and is characterized by an equatorial climate and accompanying tropical vegetation. The inland areas, on the other hand, which became known to the outside world only toward the end of the nineteenth century, have dry zones (especially as one moves northward in Kenya). The highlands consist of a series of higher plateaus and volcanic surfaces. The great African rift valley divides the central highlands into Western and Eastern highlands.

The Kenya highlands (Eastern) contain some of the most fertile lands in Africa, producing various crops, such as wheat, oats, barley, pyrethrum tea and coffee, all of which grow within a few miles of one another.[4] It is the central highlands of Kenya that the British occupied in an attempt to make Kenya into a European enclave. British contact with the tribal communities there had far-reaching economic and social changes in the lives of the people. For example, the introduction of cash crops brought about agricultural changes in Kenya, making land valuable and scarce. This ushered in significant changes in the land tenure system, introducing a movement away from communal usage of land that existed previously in traditional Kenyan society. Furthermore, the establishment of Christian missionaries to "educate" and uplift the tribal children became a vehicle of acculturation, destroying native values while exposing the tribals to the material culture and philosophy of western Europeans.

The western rift highlands, although less developed and well known than their eastern counterpart, also contain some naturally fertile areas that are densely populated. The interior plateau between the two rift systems is largely an arid and semiarid basin country except for the area bordering Lake Victoria. The divergence in land fertility led to an unequal and differential concentration of populations in the different regions of East Africa. It also triggered a pressure on land, that is, an increase in the numbers of people deriving a livelihood from land beyond what it could hold, which was further accelerated when cash crop farming was introduced by the British in the central highlands.

Economic System and Mode of Production

There are differences in the indigenous societies of East Africa, but all of them shared some early common features that derived from their common mode of production. These elements eroded with the introduction of agricultural changes geared toward a market economy. The majority of these East African people engaged early on in shifting cultivation, which suited the environmental and technological conditions of their lives. Basically, East Africa was a land of extensive agriculture. The population was limited, land plentiful and soils relatively poor.

Shifting cultivation is associated with slash and burn techniques of farming, wherein land is periodically rotated. In this type of cultivation, land must be regularly cleared of its old vegetation before it is ready for planting. This is done by slash and burn methods. A piece of land is cultivated for a few years and then allowed to remain fallow for some years. Then, a new plot of land is prepared and used for planting. This allows the original plot to regain its natural nutrients. Technology involved in this method of cultivation (hoe and digging stick) limits the size and scale of production. It also affects the economic division of labor. Production is on a small scale and is primarily for consumption. Families only hope to grow enough crops to last a season. Since there is no technology for preserving surplus, it is freely distributed when available.

An interesting aspect of this cultivation technique in East Africa was the social and economic features that resulted from it; they depart dramatically from those of people who depend upon other forms of economic activities for subsistence. Thus, horticulturists differ markedly from hunters and gatherers in their work habits and, social, political and religious organizations.[5] These differences may be attributed to those in the economic activities associated with the two modes of production. Moreover, shifting cultivation is associated with a form of social organization not shared by people who rely on more complex agricultural techniques.[6]

An interesting phenomenon of this system of shifting cultivation is the social structure that evolves in societies using this method of farming. The social system is relatively egalitarian in nature, devoid of hierarchies and stratifications visible among people engaged in more developed and complicated agricultural practices. Goody has noted that the typical African farming system (shifting cultivation) precluded the emergence of a feudal landed class. In areas where shifting cultivation was common, agriculture was carried on without the use of any animal power other than the person wielding the hoe or digging stick. There were no landlords, although, there were, of course, the "lords of the land."[7]

Africa south of the Sahara did not experience the technology of the plow previous to its contact with western Europeans. Its absence affected not only the direction of agricultural development in the region, but also inhibited the growth of social inequalities associated with farming techniques where the plow is used. The use of a plow has certain socioeconomic effects. In the first place, it increases the area of land that can be cultivated, thereby increasing its productivity. This leads to the creation of surplus, which results in differences in wealth. In the second place, it encourages fixed land holdings, allowing movement away from shifting cultivation. In the third place, it increases the value of available land and correspondingly decreases its availability. All of these features tend to contribute to a shortage of land. The absence of a plow economy therefore results in the absence of the socioeconomic consequences associated with it.[8]

The plow never penetrated precolonial Africa (south of the Sahara), since the region was untouched by the wheel.[9] The leveling of inequalities between people within these societies can be attributed to the homogenous activities that this simple technology facilitated. Distinctions arose from age, sex or kinship status. This is true of most of East Africa except the rich and fertile highland areas, which produced enough surplus to sustain complex and hierarchical societies.[10]

This egalitarianism is especially noticeable in the institution of land tenure that existed before European contact. The system of land tenure is of crucial importance in defining and shaping the power structure within a society, for it determines who may use and develop particular plots of land and, control, accede to or be denied strategic resources. In rural Africa, the control of land as a primary productive resource came to define the essential relationship of power and privilege between populations and individuals.[11] Indeed, it was in the colonial period that control over land became a basis of social inequalities in Kenya. The European colonizers owned the best parcels of land, which were used for production of cash crops. The natives were pushed off their traditional lands into "reserves" and forcibly made to work on the European farms as a way of earning money to pay the taxes levied on them by their colonial masters.

Traditional land tenure patterns in Kenya were consistent with the requirements of shifting cultivation. Land was plentiful and intertribal feuds, famines and diseases helped keep the ratio of land to people in a favorable balance. Land had no cash value, though it was crucial for livelihood. It was important for the social, political and economic life of the people. Land was considered sacred and was referred to as "the mother." Tribes had definite land boundaries within which they could cultivate or graze their animals. However, land was not collectively owned, as is commonly believed; rather, every inch of tribal territory had an owner.[12] To this extent, the concept of private ownership was not absent in Kenya. Land could be acquired through purchase or inheritance. Rights to land varied according to the relation of the individual to land. An owner did not have rights over the products of land if the producer and owner were different individuals. A person acquired the rights to the products of land through the labor spent in developing it. Ownership of land vested pride in its owner. He could not, however, exclude others from using it. Nonowners could obtain cultivating and building rights over land they did not own through kinship or residence. This may have given rise to the misconception that land was collectively owned in Kenya.

The head of the community (which varied in size) was empowered to give an individual within his group usufruct rights for purposes of growing crops for his family. The individual and his family cultivated the piece for a few years, using slash and burn techniques, and then moved on to new plots. He used land that he did not own. He could not, however, sell or abuse it. Even the owner could not dispose of land without consulting the other members of his community. Thus, land was collectively used even though it was individually owned. Pride in ownership of land also brought a sense of gratification to the owner in the thought that his land was used collectively for the good of the people.[13] Ownership of land did not give an individual economic and political power over others. There were no chiefs. Politically, the society remained egalitarian. The African groups in Kenya (except the Masai) had been highly decentralized, with councils of elders whose authority extended over a specific clan or a small local group.

This, then, was the land tenure system that existed in traditional Kenya before European contact. It allowed for the system of extensive cultivation of land based on the technology of the hoe and the digging stick and slash and burn methods. However, this institution of land tenure could not persist in this traditional form because of the agricultural changes initiated by the

Europeans in the form of new crops and techniques of cultivation, beginning in the first decade of this century. Due to their lack of interest in money,[14] East Africans were far less susceptible to the alien influence of the Westerners than were the West African agriculturists. Nevertheless, the East Africans did start to experience significant changes in their social and political life.

As noted earlier, new methods of farming and crops for cash made land a strategic resource, and forms of social hierarchy previously unknown were introduced. Relations between the owner and tenant became aggravated and there was a tendency among the tenants to hold on to land, claiming perma- nent rights to it in order to avoid possible eviction.[15] In contrast to the traditional system, the idea that the individual farmer had the right to sell or dispose of land became fully accepted. A landowning class developed because some farmers were able to acquire large parcels of land on which they planted cash crops. The wealthier farmers adopted the ox and plow; others, of more moderate means, continued to farm using the primitive traditional method of hoe and stick. The former egalitarian relationship could not survive in the light of the developing disparity between rich and poor farmers. The wealthy farmers were receptive to new ideas and techniques, and the agricultural innovations they adopted further strengthened their social and economic positions. This resulted in two farming systems: mechanized cash crop farming and traditional subsistence farming.

The commercialization of agriculture involving new methods and practices brought wide-ranging changes to Kenyan society. Agricultural innovations also upset the work schedules and division of labor between the sexes. While this happened, the old agricultural technology of the hoe continued to persist; this dichotomy contributed further to the inequalities between the classes and the sexes.

Family and Social Organization of Work

Traditionally, the ideal form of marriage in Kenya was polygyny, a union of one man to several women. Marriage was not a personal affair confined to the individuals involved; rather, it combined a series of individuals in a network of mutually reciprocal relationships. Marriage was of paramount importance to the society, for upon it rested the welfare of the entire tribe.[16] Familistic values were stressed. Generosity toward kin was emphasized and obedience and deference to elders were values that were held in high esteem.

The agricultural system of shifting cultivation and the land tenure arrangement existing under the tribal laws encouraged the institution of polygynous marriages. Not every marriage consisted of multiple wives; however, it was the ideal, desired for economic reasons. Polygyny suited the shifting cultivators, where the primary tasks of planting, weeding and harvesting the crops were deemed to be women's work. Multiple wives enabled a man to increase his assets, and in much of Africa this was one method by which agricultural expansion could, indeed, occur.[17] The economic advantage that plural wives offered a man greatly accounted for the popularity of polygyny in the regions in which it was found in Africa.[18]

Under customary laws of marriage, a man could take as many wives as he could support. A large family and homestead were desirable. In fact, a man with a large family qualified for tribal leadership; his efficient management of a large family was taken to be testimony of his leadership capabilities.[19] Since marriage entailed a transference of movable property (women exchanged for cattle, sheep or goats to compensate for her economic loss to her natal family), it was only the men with resources who could hope to marry more than one woman at a time. Men desired multiple wives for economic reasons, and only the economically well off could hope to have them.

In many instances, the first wife initiated the process of getting her husband a second and subsequent wives to help relieve her domestic burden. Women and their children were both important sources of labor in the field, each contributing their share to the subsistence economy. Co-wives helped one another in their gardens, in planting and weeding. In general, when the schedule of work was heavy, they shared tasks, including child care.

In the polygynous household, each wife was given her own hut, where she cooked, slept and kept her personal belongings. She was also allotted a plot or plots of land from the family property on which she grew food for herself and her children. The surplus was traded in the local market. She was responsible for what she produced on her allotted land and had rights over the manner in which the products of her land were used. Periodically, she was visited by her husband in her hut. Men kept separate huts of their own in which they entertained their male friends and other guests.

Clearing of the fields was, typically, man's work. Men did the intitial work involved in preparing the land for cultivation, such as felling trees and cutting and burning the underbrush. It was in such tasks that male children were particularly useful and therefore wanted for supplementing the farm work.

The other agricultural chores were performed by women. However, these work roles changed when agricultural innovations for cash crop farming were introduced. Now men were required to work in the fields.[20] As men were forced to work on agricultural fields where cash crops were grown, women were left with the responsibilities of heavy agricultural chores such as clearing the bush, which, before the Europeans came, had been exclusively within the men's domain.[21] Changes in the farming system brought significant changes in marriage patterns, family size and structure and altered the division of labor between the sexes. It also ushered in new values and economic goals.

Technological and Related Changes

East Africa entered into the exchange economy of the modern world with the arrival of the British at the end of the nineteenth century. In Kenya, this economic (agricultural) transformation brought about far-reaching changes in the way of life.[22] Kenya is primarily an agricultural country and all facets of the society are directly or indirectly dependent upon agriculture. Agricultural development took place there through the medium of colonialism, and, as a result, Kenya experienced political and social changes in its societal structure. These changes, by and large, followed upon the shifts in practices and methods of agricultural production.[23] The introduction of modern agriculture into Kenya initiated changes in subsistence methods, resulting in the growth of commerce, industry and urbanization.[24]

Shortly after World War I, European agricultural practices involving some mechanized tools, were first introduced in Kenya. Agricultural production was carried on an extensive basis on large parcels of land cleared of bush by local unskilled laborers. At the beginning of the twentieth century, beasts of burden, such as mules and oxen, were employed for light and heavy agricultural work on European farms.[25] These involved the use of plow. The acceptance of the plow resulted in the spread of cultivation of cash crops.[26]

The tractor was first used in Kenya around 1915. But it was only at the start of World War II that mechanized farming was introduced and became a part of the agricultural production process on European farms. The war created a demand for food that could only be met if machines were harnessed for agricultural production. Farming machinery was imported at this time from Canada and the United States. This period was significant for African agricultural development. On the one hand, the subsistence way of life was challenged by the introduction of a cash economy.[27] Africans were encouraged to produce beyond the needs of consumption for a market. Men's traditional roles were challenged as they were required to assume new economic roles in tune with the dictates of the time. On the other hand, dual systems of agricultural development were initiated. The modern European settler engaged in cash crop farming on the most fertile lands of the colony, while self-employed natives contined to farm through traditional means for subsistence. The modern European farms coexisted with the underdeveloped farms, where the hoe was still the essential farming tool and agriculture was undertaken principally for survival. Most Africans could not afford to farm using modern methods, because of the high cost of mechanized farming. The disparity between the two systems of agriculture was acute. According to a 1956 estimate, each settler-farmer (mainly European) owned on an average five square miles of the best agricultural land in Kenya. In contrast, the average African farmer over sixteen years of age, owned only 0.03 square miles of land.[28]

The Europeans brought with them a host of material artifacts such as agricultural machinery, wheeled transport, household utensils and other appliances. Initially, these had little effect upon the lives of the Africans. However, they could not remain indifferent or immune to the impact of the material culture of the Europeans for long. Africans exposed to the impact of these material possessions came to want them. The new possessions they acquired, brought about changes in their culture and way of life. Some of the changes were simple and easy to assimilate, such as the new variety of foods that made their way into the local diet. These new foods could be grown by the traditional method of shifting cultivation.[29] Others were more complex and difficult and necessitated radical departures from habituated modes of existence, resulting in the evolution of individualistic values. Desired material items could only be acquired through individual wages. As individual members acquired their own pay packets by selling their labor, communal ownership of material things could no longer prevail. The individual quest for material possessions, in turn, eroded the collective economy of large households. East African society and culture were under assault by the culture and technology of the incoming Europeans.

The Europeans needed labor to operate their farms. The Africans, on the other hand, did not originally need to work for wages in order to survive. They therefore had to be persuaded, in fact coerced, to work for the Europeans. The coercion came by way of taxes. A hut tax was levied on the Africans which had to be paid in cash. The imposition of the hut tax forced the African into the cash economy[30] by requiring him to either seek wage labor on European farms or earn money by growing crops for cash on subsistence farms.

Freedom of individual mobility was greatly curtailed when premium lands were taken away from the native population for permanent cultivation. Accommodations in the traditional way of life were mandated. Men were compelled to work for long stretches of time as agricultural laborers away from their homes and families. Women were left to shoulder the entire responsibility of farming, including tasks that hitherto had been assigned to men. Land acquired cash value. Ownership of land became important for the individual as a means of earning income. Successful farmers sought to buy more land in order to increase their profits. Foundations for social differentiation were laid such that the former egalitarian basis of society could not persist, not in the face of economic changes introduced by the agricultural practices and methods involving different technologies and values. Thus, while the majority of people continued to live poorly, enjoying only a minimum standard of living, the dual systems benefited some groups and individuals who succeeded in attaining status and income much above the general population.[31] The large-scale farmer operating by modern methods earned sixty times more than the average peasant who continued to farm by the traditional agricultural system.[32]

As the traditional subsistence economy was brought into contact with the modern cash economy, traditionally demarcated communal lands gave way, for farm land was no longer available to all. Hence, other viable means of livelihood and support had to be explored. Western-type education became desirable for economic success in the labor market. The school diploma or certificate became necessary for an individual to gain access to coveted bureaucratic positions in the government that carried with them lucrative, salaried remuneration.[33] Land shrinkage meant that earlier ways of living and values could not be sustained.

The shortage of land and the demand for education brought changes in basic institutional structures such as the family and the system of kinship. Polygynous marriages became less attractive as land became scarce and education for the young became valued for economic success.[34] Competition for school expenses for children fostered rivalry among co-wives, making polygyny untenable. Children lost their former economic value. They had to be groomed and trained to be economically viable. Under these circumstances, monogamous unions were preferred and familistic values were replaced by more individualistic ones. Conjugal roles were revised; the husband and wife now became the pivots of the familial structure.

In the newly emerging socioeconomic environment, the conjugal family, and not the extended polygynous family, became increasingly popular and desirable. This was, however, not without stress and conflict. Individuals were often faced with dilemmas and torn by conflicts of interest; kin loyalties and ties clashed with the desire for individual success and achievement. Women were confronted with formerly nonexistent situations. Competition for land reduced the parcels available, depriving many of their accustomed source of earning and livelihood. Women faced a loss of former autonomy and status when they were deprived of their own gardens, thereby losing their independent earnings and having to rely exclusively upon their husbands economically. They were also faced with the loss of the readily available child-care arrangements provided by polygynous marriages. Working outside the home for wages raised the question of who would take care of the children while the mother was away at work, a question working women around the world with small children must face.[35] As the Kenyan cultivators experienced changes that introduced them to the modern cash-based society, their family organizations were subject to "external pressures highly similar if not identical to those affecting western nuclear families."[36] All these conditions resulted from changes in the subsistence base of their society.

Not only were the sex roles of family members affected by the chain of events emanating from new technologies and values, but the sexes themselves were affected in ways that left women disadvantaged.[37] The complementary and cooperative division of roles that existed between men and women was challenged by new technologies introduced into the African subsistence farming by the Europeans as the Africans were plunged into a modernized agricultural system.

Before the advent of European colonization, women had enjoyed significant decision-making powers as they struggled side by side with their men for survival in the face of the harsh realities of life in East Africa. By virtue of the Europeanization of the area, however, men were funneled off to work in the mechanized farms and mines, and the women were left alone to continue the struggle on rapidly deteriorating and shrinking plots of land to produce food for their family with their traditional primitive tools of hoe and cutlass. As new technologies were introduced on large farms, women's abilities to be independent and contribute to family needs became further reduced as successful farmers (mostly males) with better machinery monopolized the markets where women had formerly sold their surplus.[38] The displacement of women from their productive positions was also partly the result of the value system of the Europeans, who conceived of sex roles along lines familiar to them. They viewed women as having the prime responsibility for child and home care and not as the producers that African women had traditionally been. In fact, when the Europeans saw the women working in the fields, while men sat around, they dubbed the men "lazy Africans."[39] It is this value system that prompted the Europeans to require men to learn new technologies and skills while excluding women, thereby contributing to the women's subordination within their previously egalitarian society. This pattern, once established, continued in the postcolonial era, even when the indigenous national governments came to power.

Conclusion

The simple technology of the hoe and stick sustained a society essentially egalitarian in its structure. Lacking the means to accumulate and build surpluses led to distribution and sharing of goods and services, with sharing and reciprocity both emphasized as important social values. These values, of course, were under attack once new technologies and the ideologies associated with them were introduced. The ox and plow and, later, farming machinery such as tractors, paved the way for burgeoning commerce, industry and urbanization. The natives were forced into a cash economy. Initially, of course, the people were unenthusiastic about earning cash, so coercion through taxes had to be exercised to get them to work on cash crop farms. However, they gradually became receptive to the positive features of a money economy. New technologies of farming brought about far-reaching changes in the socioeconomic structure of Kenyan society. For one thing, it led to the exploitation of poor farmers, resulting in widespread underdevelopment; only a few rich and wealthy farmers benefited from the transformation of the economy. Family organization changed to suit the demands of a modern cash economy. Familial values gave way to individual ones. Sex roles were affected, with women especially having their status affected by the new technologies. Within a short span of a few years, Kenyan society ceased being only subsistence-based. It entered the market economy. This suggests that certain technologies eventually lead to inevitable structural and societal value changes. In the case of Kenya, the advent of the plow and later the tractor and other farm equipment ushered in a shortage of land, thereby breaking the traditional "communal" structure of land tenure. Paradoxically, introduction of this new farm technology created both scarcity and surplus at the same time, thus encouraging competition for strategic resources among the population. Competition replaced sharing. This was unavoidable. Under such changed agricultural conditions, brought about as a consequence of the introduction of modern technologies by the Europeans, former egalitarian structures and values sustained under hoe cultivation were extinguished in Kenya.

Notes

1. Ernestine Friedl, *Women and Men: An Anthropologist's View*, New York: Holt Rinehart & Winston, 1975, p. 47.
2. Ester Boserup, *Women's Role in Economic Development*, New York: St. Martin's Press, 1970, pp. 16-24.
3. W. P. Lineberry, ed., *East Africa*, New York: H. W. Wilson Company, 1968, p. 22.
4. Lineberry, p. 22.
5. Friedl, p. 46.
6. Boserup, pp. 15-24.
7. Jack Goody, "Economy and Feudalism in Africa," *The Economic History Review*, Vol. XXII, No. 3, December 1969, pp. 396-397.
8. Goody, pp. 396-397.
9. Goody, pp. 396-397.
10. C. G. Widstrand, *Cooperatives and Rural Development in East Africa*, Uppsala: Scandinavian Institute of African Studies, 1970, p. 43.
11. Ann Siedman, *Planning for Development in East Africa*, New York: Praeger Publishers, 1973, p. 159.
12. Jomo Kenyatta, *Facing Mount Kenya*, London: Secker and Warburg, 1953, p. 2.
13. Kenyatta, p. 26.
14. Melville J. Herskovits, *The Human Factor in Changing Africa*, New York: Alfred A. Knopf, 1967, p. 379.
15. Robert L. Tignor, *The Colonial Transformation of Kenya*, Princeton, N.J.: Princeton University Press, 1976, p. 307.
16. Kenyatta, p. 165.
17. A. Richards, *Economic Development and Tribal Changes*, Cambridge, England: East African Institute of Social Research, 1952, p. 204.
18. Boserup, p. 37.
19. Kenyatta, p. 175.

20. Tignor, p. 304.

21. Tignor, p. 304.

22. L. W. Cone & J. F. Lipscomb, eds., *The History of Kenyan Agriculture*, University Press of Africa, 1972, p. 13.

23. Cone & Lipscomb, p. 23.

24. Beatrice B. Whiting, "Changing Family Style in Kenya," in *The Family*, Alice Rossi et al., eds., New York: W. W. Norton and Company, Inc., 1978, p. 211.

25. K. R. M. Anthony & B. F. Johnson, et al., *Agricultural Change in Tropical Africa*, Ithaca, N.Y.: Cornell University Press, 1979, p. 120.

26. M. E. Luckham, "The Early History of the Kenya Department of Agriculture," *East African Agricultural Journal*, Vol. XXV, No. 2, October 1959, pp. 97-105.

27. Cone & Lipscomb, p. 27.

28. Ann Siedman, *Comparative Development Strategies in East Africa*, Nairobi: East African Publishing House, 1972, p. 5.

19. Cone & Lipscomb, p. 27.

30 Siedman, *Comparative Development Strategies*, pp. 28-29.

31. Siedman, Comparative Development Strategies, p. 49.

32. Siedman, Comparative Development Strategies, p. 53.

33. Whiting, p. 217.

34. Whiting, p. 217.

35. See Whiting, pp. 211-224, for an excellent analysis of the changes in the Kenyan family.

36. Susan Abbot, "Full-Time Farmers and Week-End Wives: An Analysis of Altering Conjugal Roles," *Journal of Marriage and Family*, Vol. 38, No. 1, February 1976, p. 72.

37. Ann Siedman, "Women and the Development of "Underdevelopment:" The African Experience," in *Women and Technological Change in Developing Countries*, Roslyn Dauber and Melinda L. Cain, eds., Boulder: Westview Press Inc., 1981, p. 111.

38. Siedman, *Women and Development*, p. 115.

39. Boserup, p. 19.

CHAPTER 5

Plow Cultivation:
The Indian Example

Introduction

For centuries, the Indian system of agriculture remained unchanged. Agricultural tools and techniques, hundreds of years old, have persisted well into the twenty-first century. A light plow drawn by bullocks was, and still remains, the most important agricultural implement for vast majorities of cultivators.[1] Oxen or male water buffalo are the principal source of traction for plowing Indian fields.[2] These animals are to the Indian farmer what the tractor is to his American counterpart. For the Indian peasant, the cow is an essential animal on the farm. It is a source of energy and food (non-meat). In fact, the survival of the farmer depends upon the cow. The system of cultivation using draft animal and plow has persisted for centuries, and only recently have certain technological changes been initiated. The study of this changing technology reveals a fascinating example of the complex interrelationship between the human and technological forces in the process of change. The forces that prevented the Indian agricultural system's receptivity to technological innovations in the past, and the factors that have allowed these inhibitions to break down in more recent years, are described below.

Physical Features

India is a land of great geographical and climatic variety with contrasts far greater than can be found in the continental United States. Physically, it is cut off from the greater part of Asia by the mountain ranges of the Himalayas, which extend across northern India from west to east in an irregular crescent. The Himalayas, partly in India and partly beyond India's northern boundary,

Map of India showing Punjab

are responsible for isolating the country from two other mainstreams of Asian civilization: China and the Islamic world of the Middle East. The mountains form a natural barrier; it is only through passes in the northwest that invaders and immigrants came to India, only to be absorbed into the cultural matrix of Indian life.

Directly south of the Himalayan mountains lie two great river valleys. India's holiest river, the Ganges, flows southeast, parallel with the mountains, through one of the most fertile agricultural areas in the world. The Indus, is Pakistan's chief river; it flows southwest and south through drier lands. The Indo-Gangetic plain, which is 1500 miles long and 150 to 200 miles wide, is formed by the basin of three great rivers of the Indian subcontinent; namely, the Indus, Ganges and the Brahmaputra. This plain represents the most extensively cultivated and densely populated area particularly where water is available. This includes the region of Punjab in northern India, the area that will be the focus of our attention.

The State of Punjab in India lies in the northwest of the Indian subcontinent. It was carved out of the region known as Punjab in undivided India at the time of its partition in 1947 when India was divided into two nation states based on religion. West Punjab became a part of Pakistan and East Punjab became the State of Punjab in India. Pakistan borders the Indian State of Punjab on the west; on its north lie the Himalayan Mountains and the state of Kashmir; the desert of Rajasthan surrounds the south. Punjab, a part of the great Indo-Gangetic plain, contains a fertile plateau with some of the most productive land in the world. It is the granary of India and has been compared to the agricultural region of the Midwest in the United States.[3] In fact its yields per hectare are the highest in world.[4] This has been made possible by the technological developments that have taken place there recently. Interestingly, the Punjab stands as a dramatic example of agricultural development not shared by other Indian regions and therefore provides an excellent case study for understanding the human and social factors crucial in technological change.

Agricultural Mode of Production

The system of agriculture in India is still largely traditional. Only a small number of farmers have begun utilizing modern agricultural equipment such as tractors with several different attachments for performing variegated tasks such as plowing and cutting the grain. These farmers, with large land holdings, are prosperous and are able to afford the luxury of using modern farm machinery for cultivation. For the average farmer with smaller land holdings, the cost of modern farm machinery is beyond his means. Consequently, the most common agricultural tool available to him is the wooden plow, tipped with half-inch steel and pulled by a pair of bullocks. The wooden plows outnumber iron ones by a factor of more than thirty to one.[5] Even bullocks and plows are not available for ownership to all farmers. The steel plow, taken for granted in the Western world, is not common farm equipment. Rather, the only agricultural implement for many peasants is a heavy, short-handed hoe used for hoeing and digging. There is also a shortage of traction animals for plowing. For each farm of ten acres or less (average land holding for a small farm), one pair of oxen or bullocks is considered adequate. There are sixty million farms in India but only eighty million traction animals, leaving a deficit of forty million.[6]

The draft animals on the farm not only provide the necessary energy for traction to the ordinary peasant (in the absence of a tractor, an expensive item), they also serve other functions comparable to those served by the petrochemical industry in the more developed countries. The cattle of India provide 700 million tons of recoverable manure, half of which is used to fertilize the fields; the rest is used as domestic fuel for cooking. Cow dung, therefore, is an immensely important source of energy for the farm family.

India's agricultural production is still susceptible to the vagaries of nature. Only one-fourth of the net farm area is irrigated and less than two-thirds of the potential available area can be irrigated from surface and groundwater resources.[7] Bold irrigation systems, extending into India's historical past, have been recorded. Independent India has pushed ahead with massive irrigation projects. Yet it is only in the past three decades that agricultural production has begun to benefit from irrigation schemes due to the socio-economic and technological forces that have come into play. Punjab, primarily a wheat-producing region, has shown a remarkable increase in agricultural production and has been heralded as leading India's green revolution, a term referring to phenomenal increases in agricultural development due to techno-logical achievements. The Punjab in India considered India's bread basket, is one of the fastest-growing agricultural regions in the world and, as such, its case study should be of value in itself. The study of the agricultural develop-ment of the Punjab offers an added bonus in providing an excellent opportu-nity to study the transition from traditional to modern agricultural practices. The social history of agriculture in the Punjab is discussed below to provide a background for understanding the complex interrelationship of human and structural forces crucial in technological change.

Social History of the Agricultural System in the Punjab

The system of cultivation using the plow has been characteristic of agriculture in the Punjab dating back to Mogul times in the sixteenth century. Under the Moguls, the peasants labored under a system of tax farming. They provided the revenue for the Mogul Empire. It is through their produce that the standing army of the Mogul Empire was maintained. Specific areas were assigned to individual tax collectors responsible for assessing and collecting revenue from peasants. The peasant had little incentive to increase his production, since the surplus was siphoned off as revenue by the tax collector for his personal use or for the imperial treasury. In fact, this system may have acted as a disincentive for the peasantry, resulting in indifference and reluctance to grow more.[8] As a consequence, the agrarian order remained unchanged and technologically stagnant.[9]

After the decline of Mogul rule, the Sikhs rose to power in the Punjab. Even so, agriculture did not improve.[10] The system of farming under the Sikh rulers did not make agriculture profitable for the tillers of land. The state or the revenue officers took as much as they could, leaving the cultivator with barely enough to subsist on. The land tax was paid not in money but in kind, such as crop produced, and was proportionate to the produce of the field. In these circumstances, after the revenue was skimmed by the state, there was no profit that accrued to the peasant cultivator.[11]

The excessive revenue demands and the low cash returns on the crops made possession of land an encumbrance rather than a privilege. Often, excessive revenue demands made payments prohibitive. The cultivators would abandon their lands and move to other parts of the province, very much like the peasants in Mogul times. Essentially, the state of agriculture prevalent under the Moguls persisted even though the political power had shifted to the native population. Agriculture remained unprofitable and provided a poor basis for investment because of the poor returns derived from it. When the British took over control of the Punjab in 1849, they inherited an agricultural system that was basically inefficient and technologically primitive.

The British initiated a series of changes relevant to the system of agriculture in the region. The most direct impact of British rule was on land revenue policies.[12] The inauguration of British land policies stimulated a high turnover

Bullock-drawn plow used to prepare a field Artist: Swapna Das

of land. Consequently, the position of many individuals rose and fell. However, British land policies affected only individuals; change was more often in individual status rather than in the substance of agrarian society viewed by social and caste group.[13] Nevertheless, the impact of British rule in the Punjab was not inconsequential as far as the agricultural classes were concerned. The British did not act on a passive, agrarian society. This is evident from the responses of the cultivating class to some of the measures initiated by the new rulers. In fact, the British, too, recognized this dynamism and reduced agricultural taxes in the hope of bolstering their revenue through the creation of new taxable holdings.

As the British consolidated their power in the Punjab, they introduced a series of measures for assessment and collection of taxes from the countryside. These measures had far-reaching consequences on the land tenure arrangements and the cultivating classes that lived by them. The British rule brought law and order. It provided security in tenure, and land became a valuable commodity. The province of Punjab advanced rapidly in prosperity. With rise in prosperity the agriculturists found themselves in possession of valuable property such that they had not known before.[14] British rule brought agricultural prosperity by bringing good administration, a settled land revenue and a tremendous increase in canal irrigation.[15] It also brought increased indebtedness for the Punjabi peasant. Indeed, the conditions responsible for prosperity contributed to the increased indebtedness of the peasants. The capital from surplus available for borrowing and its abuse by usurers, famines, rigidity of land revenue collection and new civil laws favored the moneylender rather than the ignorant cultivator. All these factors contributed to the impoverishment of the peasants.[16] Thus, it has been suggested based upon the historical evidence that the reforms of the British did not benefit the cultivating class.[17]

Nonetheless, the British established individual property rights over the land. In pre-British Punjab, the peasants of the village had collectively owned land. Land was inalienable. The British, however, believed that the peasant happily tilling his plot was conducive to a peaceful British rule.[18] Thus, where land ownership in the village was uncertain, it was given to any peasants who could show that they had been its cultivators for twenty years. This decision was significant, for it produced thousands of new landowners with rights of sale, mortgage and hereditary transfer. The peasant who cultivated the land became its owner or proprietor. This ushered in inheritance. For the first time, land became a marketable commodity and could be sold and purchased like any other goods.

While the British granted tenants proprietary rights, they also inadvertently contributed to increasing the value of agricultural property. By bringing peace to the region, introducing an orderly system of tax collection and keeping revenue assessment low, they made agriculture profitable. Surplus, which could be sold for revenue, was now possible. Construction of better roads and railways facilitated the transport of agricultural surplus to markets. New markets opened up for the sale of surplus produce. Agriculture could therefore yield profits beyond mere subsistence. This meant that people with means could profitably invest in land and expect returns from it. Ownership of agricultural property became important. It gained immense value. In 1886, land sold for ten rupees an acre; by 1921-26 the price had skyrocketed to 238 rupees an acre. These changes in land ownership and value increased the role of the moneylenders, who were not a powerful group on the rural scene until the British arrived.[19]

Money lending existed even prior to British rule, but was less common. The vagaries of nature made it essential for the peasants to borrow from the moneylender to meet their survival needs. Two factors mitigated the power of the moneylender. First, a strong village community that supported the cultivator, thereby making unreasonable extraction impossible and kept him in check. Second, state apathy regarding recovery of money loaned[20] led to informal settlements affordable to the cultivator. The decline of the village community and establishment of civil courts under British rule introduced the reign of the moneylender.

The British imposed a land tax known as a land revenue system. This represented a fixed portion of the gross produced by the peasant, which had to be paid in cash. Under the British, taxes were lower compared to those demanded by the Sikhs, but they were collected more rigorously and demanded in cash. The Sikh revenue collection had been made in kind and the inability to pay taxes affected not the cultivator but the rich district officers responsible for tax collection. The cultivator did not have to surrender his land to the moneylender in order to pay taxes. Under the British, however, the payment of revenue was the peasant's responsibility. This was not always easy. Inability to save sufficient cash reserves for payment of taxes or uncooperative weather often forced the peasant cultivator to turn to the moneylender for help. The moneylender benefited tremendously from the land revenue system introduced by the colonial government. In fact, the primary impact of British rule in the Punjab was the strengthening of the position of the moneylender in the village economy.[21] Conditions very favorable to the money-lending class were created when land gained in value and became a source of wealth. Investment in land became lucrative. Money lending became the most profitable business in the region. In 1868, according to the Punjab census, there were 53,263 bankers and moneylenders. By 1911, their numbers had grown to 193,890. The domination of moneylenders in the Punjab was unrivaled by any other Indian state within the British Empire. In fact, one-fourth of all moneylenders lived in the Punjab.[22] They did not all belong to one caste. They were either Brahmins or members of the commercial castes. They were, however, all members of the upper stratum of society.

The moneylender was not interested in owning land in itself, for such ownership yielded little profit. Rather, his interest lay in what the land produced. He was interested in being a mortgagee of land. He realized that the commercial value of land was derived from what it yielded. As soon as markets became available within easy reach, the moneylender set out to establish control over land by gaining control over the cultivators. In being a mortgagee, the moneylender got the benefit of the land plus the power over the peasant, who was the owner-cultivator. The profits that accrued to him from this arrangement are evident from figures available from the Punjab on the proportion of cultivators who had their land mortgaged to the village money-lender. The moneylenders became a parasitic stratum feeding off the lifeblood of the peasant cultivators. Under such circumstances the cultivating class had little incentive to innovate, knowing that any increase in production would be skimmed off by the moneylender.[23] For peasants the situation in the Punjab, although different from that which prevailed under the Moguls or the Sikhs, failed to provide any incentives that would promote technological change in agriculture. Under the British, the rights of ownership were endowed to the individual, allowing inheritance to occur. However, the peasant's inability to accumulate or retain wealth made any kind of innovative technological investment in their land, impractical and useless. Thus, the centuries-old simple agricultural technology persisted. The primitive method of cultivation that persisted was not due to the lack of abilities of the peasant cultivating class. On the contrary, scholars of Indian agriculture, have noted the wisdom of Indian peasants and what they can teach the western agricultural experts.[24] In 1889 an agricultural expert Dr. Voelcker, had noted that 'there is little or nothing that could be improved'[25]. The lack of productivity of Indian agriculture was due to lack of systemic resources available to the peasant farmer, not scarcity of knowledge. The above discussion underscores the importance of structural features of society that can inhibit or be conducive to technological changes.

Social Structure and Village Economy

At the time India gained its independence from the British Empire in 1947, the Punjabi village remained economically self-sufficient, producing very little surplus. The village economy was based on land, and most households derived their livelihood either directly or indirectly from it. Socially, the village was stratified along caste lines. This was true of all members of the rural community, regardless of religious affiliation. Even the Sikhs were not free from the impact of caste ideology, although Sikhs overtly reject the principles of the caste system.

Caste was a mechanism of organizing social relationships between individuals and groups. The idea of ritual pollution, purity and mutual exclusion provided the basis for inter-group relationships between castes. At the very top of the social hierarchy were the Brahmins. The lowest rung of the society consisted of the various castes generally known as the untouchables. Between these two polar ranks were interposed a variety of castes both high and low, "clean" and "unclean." The pattern of caste dominance that existed under the pre-modern economy has since changed. In present-day Punjab, a universally agreed-upon system of hierarchical ranking is absent, although the ritual purity of Brahmins and the political superiority of Jats are acknowledged.[26] Traditionally Jats have been cultivators and by and large continue to be so. Agricultural changes contributing to the welfare of the cultivating class, mainly Jat Sikhs, broke the hegemony of the Brahmins and established the Jat Sikhs' dominance in the Punjabi society. Governmental support through agricultural reforms that increased the farmers' ability to improve agricultural production and profits provided them the opportunity to be economically independent. Economic independence allowed cultivators to be socially free. Henceforth, they could follow and enforce their beliefs and practices rather than those of the Brahmins, who until recently had been the dominant group in society.

The pre-modern village community consisted of two main occupational divisions. There were those who worked on land as peasant cultivators or farm laborers and others involved in various "service" occupations. The latter included carpenters, tanners, sweepers, water carriers, barbers and blacksmiths. These castes were hierarchically arranged in terms of high and low social status; differential rights and privileges emanated from caste membership. Caste legitimized inequality, and members of a given caste knew the behavioral norms and expectations accompanying their caste status. One's occupation in life was traditionally determined by birth, which in practice meant caste. This is now changing with the expansion of the agricultural economy and surrounding industrial urban growth.

Before agricultural mechanization in the Punjab, economic relationships between castes were governed by the *jajmani* system, a mechanism that facilitated the exchange of goods and services between castes on a hereditary basis. Different castes were interlocked into a patron-client relationship, each providing the other goods and services. However, the relationship was not equitable between the groups involved in this economic exchange. Landowning cultivators were the patrons who employed the "service" castes, their clients. Payments by the patrons for the goods and services received from their clients were not made in cash but in kind and were spread throughout the year. Thus, for example, a Jat farmer, the patron, would pay on a regular basis a portion of the grain he produced to the barbers or water carriers for the services they provided his household during the year. This relationship continued over generations. Hence, the death of an individual patron or client did not terminate the patron-client relationship between the respective families.

In a technologically simple pre-modern village community, the *jajmani* system worked well. There was little mobility, and caste members were assigned their social and occupational status by tradition. Occupational specialization according to caste further eliminated competition for jobs. The castes were tied into an inter-dependant relationship, each caste contributing its part for the maintenance and continuity of the whole system. The caste system functioned in ways similar to an organism where different parts of the body of the organism maintain their activities such that the whole (body) continues to persist. Castes protected their members against wrongdoing by others. The system offered security and stability for all its members.

The family was patriarchal and extended, organized around lineal relatives related through common male link. Descent was patrilineal with emphasis on the male heirs to continue the family line. Boys were desired and the birth of a son was an occasion of joy. Upon marriage the men brought their wives to live with them in their parental household in their village. The girls upon marriage became part of their husband's family and kin group. This structure of the family suited the agricultural needs of the society. Given the low level of technology, family members provided the much-needed labor on the farm. This situation was comparable to pre-industrial Western Europe, where family members were tied into a relationship of economic interdependence. In the

Punjab, sons were important for economic reasons. Their labor was indispensable for family survival and its economic well being. A farmer without sons was like a man with no hands. Large families with many children, especially sons, were desirable. Daughters were considered transient members of the family similar to the traditional Chinese family. Upon marriage they became part of their husband's parental households. Hence, boys were preferred to girls.

Since the average peasant owned small sized plots of land, inheritance could further fragment the land holdings. Sons often remained together in order to earn a living from the land. Indeed, the need for brothers to stay together was especially great among the class of farmers who had limited resources and could ill afford the expense of hired labor or labor saving agricultural machinery. Farming was man's work, and men performed the tasks associated with cultivation, such as preparing the fields, plowing and harvesting. A farmer's wife tended to household chores that freed the men to work on land from dusk to dawn and, thus devote their undivided attention to farming. Women on the farm had supporting roles. The exceptions were members of the lower castes; these women worked alongside their men in the fields as agricultural cultivators. The family and caste were the organizational backbone around which the village community was woven providing its members with the essentials for survival and stability as they went about the business of living.

Thus, the village remained, until recently, relatively isolated and autonomous. Agriculture continued as the main source of livelihood for most people and agricultural tools and equipment remained undeveloped and traditional for centuries. This situation began to change when social and economic conditions began to undergo a transition, making technological innovations possible. Material aspects of life for the cultivating class were altered by the changes initiated in the societal base. These involved curbing the moneylenders' power and making attractive loans available to the farmers for agricultural improvements. Through the liberal credit policies and land tenure reforms adopted by the government of independent India, the farmer was retrieved from the clutches of a stratum of society, namely, the moneylenders, who under colonial rule had dominated the rural scene.

Social Factors and Technological Change

The British, concerned with the exploitative role of the money lending class in the Punjab society, initiated reforms[27] to halt the process. However, failure to deal with the circumstances that contributed to the impoverishment of the cultivators allowed the moneylenders' hegemony to continue. An initial blow to the money lending class wad made by the government of Punjab in 1937 when debts owed to money lenders where interest payments had doubled the principal amount were canceled. A more serious directive later accompanied this in 1949, which provided loans to cultivators without debt at interest rates low compared to what the moneylenders charged. The provisions for alternative credit combined with the growth of cooperative societies were important developments that affected cultivators. However the resources for these provisions had to be generated mainly by the farmers, who were required to put their savings in credit societies. Therefore these changes although meant to benefit the cultivators, still failed to provide them the necessary incentives. The real consequences of these developments in changing the conditions of life for peasant cultivators came much later, in the decade of the 1960s, when the cooperative movement gained momentum through governmental support. Consequently, rural credit for agricultural development became easily available to eligible farmers.

At the time of the independence of India in 1947, the Punjab basically had a subsistence-type of agriculture. Small amounts of cash crops were grown to pay off land revenue tax and meet family cash expenses.[28] Agricultural development and technological changes occurred only after the government of India and the state government of Punjab adopted a policy of community development programs to help the villagers help themselves. These programs provided a solid economic environment in which technological changes became possible, ushering in the transition from traditional to modern agricultural methods and practices; this included agrarian reforms, which became particularly useful in giving the Punjab a head start with respect to other Indian states. Of special importance in the reforms that took place were tenancy reforms that helped the agricultural class establish a pattern of class relationships that promoted the ascendancy of the progressive farmer.[29] As a result, the landowning Jat Sikh farmers became dominant. Thus, the alteration of the patterns of class relations between groups played an important part in the technological changes that occurred in the Punjab, making it the most advanced agricultural region in the Indian union. The factors responsible for this are described below.

Technological breakthroughs occurred in the Punjab in the 1960s. As part of overall agricultural development, measures relative to land tenure and land holdings were initiated in the Punjab. These gave Punjab an edge over other states, where various forms of tenancy had inhibited the adoption of modern agricultural methods of farming.[30] In the Punjab in 1961, 52 percent of all farmers were owner-cultivators. By the 1970s tenancy was uncommon among farmers. Most of the cultivators were now peasant proprietors operating their own land. Now only 9 percent of the peasants were tenants and accounted for only 10 percent of the total cultivated area in the state. As far as tenancy and operational holdings were concerned, the region was fairly uniform.[31]

In 1960–61, as part of a nationwide pilot program, the district of Ludhiana in Punjab was selected for introduction of a new agricultural strategy popularly known as the "Intensive Agricultural Development Program" (IADP). The IADP imported a model of agricultural strategy that was seen as freeing the Indian system of agriculture from the 'shackles of the past'. Primarily American experts were involved in the introduction of IADP in the Punjab through several international agencies. These included some private American Foundations (Rockefeller and Ford Foundations), the American Government and such international agencies as the World Bank and the USAID. Essentially these agreed to provide the credit and financial assistance that would be needed to implement the new agricultural policies advocated by them. The emphasis under this strategy/program was not to provide piecemeal help. Rather, its approach was for overall agricultural development by strengthening the economic basis of the society through a strong financial structure of cooperatives whereby timely and adequate help for farmers could be available when needed. These cooperatives loaned money to cultivators provided they owned land and paid their dues. The loans could be used to buy agricultural implements such as pumps, tractors and fertilizers, as well as to pay back any past debts. This measure had wide-ranging effects on the social structure. By making loans available that could be used to repay past debts, the cooperatives dramatically contributed to the demise of the moneylender's power and domination over Punjabi society. The Jat cultivators now became financially independent.

Also pivotal in promoting technological innovation was the program of land consolidation that was first started in the 1930s. Small land holdings did not favor mechanized farm equipment such as tractors to boost production. Such expenses could not be justified; neither was it economically feasible for the cultivator of such farms to buy this equipment. The land consolidation program was based on the belief that land broken into small parcels, due to inheritance, becomes uneconomical because of the space wasted in boundary markings and time lost in moving back and forth between fields. Pooling land to promote efficiency and thus gain better returns could eliminate this waste. The procedure involved combining all land on paper and dividing it among the cultivators, with each person receiving "a share equal to his share of the original total amount."[32] After consolidation, the plot allotted to a person would be in an area where he held his largest fragments. Thus, each farmer, instead of having several fragmented parcels, had one continuous piece of land.

By 1968-69, almost all the agricultural land in the Punjab had been consolidated; this made it possible for farmers to use some modern farm equipment such as tube wells and tractors, which resulted in increased crop production. Therefore reforms introduced that affected the system of land tenure and land consolidation partially contributed to the adoption of new agricultural technology.[33]

The size of the land holdings was crucial in the adoption of the new mechanized farming.[34] In many villages of the Punjab, some elements of new technology had been introduced earlier in the century; however, their spread did not occur until the middle of the century. Small plots of land were a crucial constraint inhibiting mechanization. In Ludhiana, for example, the replacement of bullocks by tractors for landholdings of twenty-five acres or less was considered uneconomical.[35] Mechanization was not feasible; accordingly, the family had to rely on its own labor for agricultural productivity. Indeed, the more family members there were to contribute, the better it was. When the productivity and thus the income of a peasant family increased significantly as a result of their numbers, the savings, in turn, were used to acquire more land and bring more acreage into cultivation. Only then did mechanization become possible.[36] Hence, mechanization occurred only after the farm family achieved a level of economic well being that allowed for departure from the traditional means of farming. This indicates the significance of socioeconomic factors (in this case size of holdings) in the success of technology. The relationship of technological development and socioeconomic structure is like a two-way street. Each is important in providing an incentive for the other.

Once mechanization entered into the farming operation, it brought about other significant technological changes. Mechanization promoted the adop-

tion of new, high-yielding varieties (HYV) of seeds that ushered in the Green
Revolution.

> It was precisely in the areas where tractors and irrigation facilities were already
> concentrated on the eve of the introduction of HYV's that the farmers promptly and
> massively adopted the new technology of high-yielding varieties of seeds.[37]

The crucial factors in the agrarian structure that promoted technological
change in the Punjab were noticeably absent in other areas where the green
revolution failed to occur.[38] These factors included (1) decline in tenancy, (2)
absence of wage labor, (3) increase in middle-sized land holdings and (4)
irrigated land areas. All of these conditions are indicators that the peasants
were owner-cultivators who had achieved a certain level of economic well
being and were able to cultivate land where water supply was assured. Thus, it
is not an accident that the Green Revolution occurred in prosperous regions
and failed to occur in other areas. It was economic prosperity that promoted
the adoption of this new technology of tractors and high-yielding variety of
seeds. One need only recall that under conditions of pauperization, the
Punjabi peasant had been reluctant to innovate.

The new high-yielding variety of seeds substantially increased productivity
of all land-holdings, including small parcels. This technology gave further
impetus to agricultural modernization in the Punjab, and, in turn, acted as a
catalyst for agricultural innovation.

> The introduction of high yielding varieties of wheat stimulated private investment in
> land improvement and farm mechanization and thus set in motion the process of self
> sustained capitalistic growth.[39]

Of course, Punjabi agriculture, with its large and well-irrigated
peasant-operated farms, could easily absorb this new technology.[40]
Accordingly, the socioeconomic features of the agricultural society proved
important in fostering technological change.

Jat Sikh farmers' receptivity to technological innovation has also been recognized. As a group, the Jat farmers of the Punjab had been exposed to
modern values. Many have served in the armed forces and, after retirement
from the army, have become farmers. Others, having traveled outside India
and worked in foreign countries for many years, brought their savings back to
the Punjab and started their own family farms. Furthermore, the literacy rate
among the Jat Sikhs was high. It too, undoubtedly must have facilitated
exposure to modern ideas and values. Above all, the Sikh community is
known to be hardworking, enterprising and independent in spirit. All these

factors combined with the social conditions noted earlier may have been conducive to acceptance of new technology to foster agricultural change in Punjab.

Once technological change occurred, it also initiated a series of changes in the social system. Adoption of a given technology is contingent upon the availability of an appropriate social base. The removal of social obstacles and the provision for adequate incentives are important in the acceptance of technological change. However, once such change occurs, it in turn initiates changes in the social system. Thus, the relationship between technological and social change is reciprocal.

Technological Change and Social Consequences

Mechanization of agriculture has increased in recent years in the Punjab. This is indicated by the numbers of tractors per cultivated area, and, more importantly, by the increase in the absolute number of tractors. According to the 1971 census, there were 10,636 tractors in use in the Punjab out of 53,334 for the nation as a whole. Twenty percent of all tractors in India are in the Punjab. Hence, the Punjab leads all other states of India in the number of tractors in use. There are 2.8 tractors for every one thousand hectares of net sown area. In the Punjab, as Ahmed points out,

> the study of a cross section of Indian states reveals that the areas in which large proportion of farmers adopted HYV wheat, were precisely those where a large number of tractors were already in use.[41]

In Punjab there was also an increase in the use of chemical fertilizers and pesticides and land under irrigation for agricultural production. Compared to the rest of India the use of fertilizer per hectare (2.47 acre) of land was three times the national average. Similarly 80% of the land in Punjab was under irrigation. The comparable figures for rest of India was approximately 30% according 1981 census data.

The adoption of the new high-yielding varieties (HYV)of seeds made available in the 1960s, part of the intensive agricultural strategy of the Green Revolution brought about agricultural prosperity. But his agricultural prosperity involved social and other costs not anticipated when the technology was first adopted. These consequences were significant socially and politically for the Punjab in particular and the nation in general.

The traditional Punjabi village was, by and large, a self-sufficient community. It maintained its equilibrium with various castes tied into a pattern of economic exchange based on caste norms and traditions. Agricultural mechanization upset this equilibrium by disturbing the nature of obligations between the different castes. It also displaced human labor. This is exemplified by the use of tube wells and tractors, both of which displaced labor. Before tube wells, there had been Persian wheels, and, before them, water- lifting buckets, which had been used for supplying, water for crops. The bucket used for lifting water from the wells or ditches was made of iron or leather. It was suspended by a rope at the end of a pole and lifted water by a seesaw motion.[42] It was manually operated and was used by peasants on holdings that were too small for the economical use of bullocks.

Persian wheel Artist: Swapna Das

The Persian wheel necessitated the use of both human beings and draft animals. One person drove the bullocks and another controlled the water wheel.[43] However, human labor partially could be dispensed with if the animals were blindfolded so that they did not go astray from their track and the person controlling the water was near the well and occasionally shouted at the animals. In the absence of draft animals, several children had to perform the task.[44] Both the water-lifting bucket and the Persian wheel required human and/or animal power. However, the Persian wheel required more power than the water-lifting bucket. Even as late as the mid-1950s, only a minority of farmers had Persian wheels. An improved version of the Persian wheels, which had attached pumps to draw the water, was introduced into the market but not popular because of its higher cost. These tools, simple as they were, required human supervision. The mechanized tube well ultimately replaced the Persian wheel by the 1970s.

The tube well used today consists of a pipe attached to a strainer, inserted into the soil, through which water is lifted by pump. It requires very little supervision and will pump water an entire day. Water lifting has been one of the most dramatic areas of technological change. The tube well freed the labor that was formerly required for field irrigation; that freed labor could now be used elsewhere. The tube well also increased the quantity of water available for irrigation. It was more efficient. But the more important impact of this changing agricultural technology was the transition of traditional agriculture to supposedly a modern, scientific one. A new concept of time and labor were in the offing, for both involved expenditure of money and had to be monitored.

Agricultural tools have also become increasingly mechanized. Traditionally the plow driven by draft animals was used to break up the earth and prepare the fields for sowing. An average pair of bullocks could plow an acre in eight hours.[45] This was recognized as slow. The emphasis on increasing plowing speed ushered in the use of tractors.[46] A fifty horse-power machine can plow an area in half an hour that formerly took a team of bullocks a day. In addition, one person is sufficient to oversee several farm operations. This means that hired labor formerly used to prepare and sow the fields can be dismissed, thereby increasing the earnings of the farmer. In Ludhiana alone, by 1970, there were as many as five thousand tractors in use. In comparison, in the 1940s, the number of tractors in use was negligible.

The wealthy farmers who mechanized their farm equipment benefited enormously However, the fruits of mechanization were not equitably shared. For example, as George noted, on one large farm in the Punjab "nearly ten thousand farmers and government officials gathered to watch two John Deere self-propelled combines demonstrate the practicality of mechanical harvest-ing."[47] They were amazed by what they saw. What had taken three or four men an entire day to accomplish was now done within an hour. "Hundreds of self propelled combines are used in productive wheat growing regions" and these have "eliminated the dependence on a large number of farm workers at harvest time."[48]

While mechanization has benefited the rich farmer in dispensing with field laborers and the problems their employment could create, it also dis-placed the labor that had been guaranteed a source of livelihood under the traditional *jajmani* system. Many of the traditional payments in kind were replaced by cash payments as a result of the rising prices of food grain. It became economical for the farmer to pay his workers in money rather than in grain. The most serious effect of mechanization, according to Frankel, has been the "rapid deterioration in the good relations' between the land-owners and agricultural laborers in the context of an accelerated erosion of traditional ties based on payment in kind."[49] Relations based on mutual obligations among different groups of people involved in the village economy that had traditionally persisted for centuries succumbed to market forces leading to disastrous changes that reverberated far beyond the boundaries of the State of Punjab (discussed in the next section). Farm mechanization led to increases in inequality and poverty by depressing wages of agricultural laborers and lowering grain prices. This made earning a living even more difficult for small owner-cultivators.[50] These kinds of technological changes have led to the development of larger-sized farms and consequently to even greater concentra-tion of the land in the hands of a few.[51]

The benefits of farm mechanization accrued principally to the wealthy farmers. All farmers were able to take advantage of the opportunities offered by the introduction of the Intensive Agricultural Development Program. However the farmers with holdings of twenty acres or more made the greatest gains.[52] Therefore, a serious disparity emerged between these large farmers and the majority of other cultivators.

The financial gains netted by farmers with holdings of twenty or thirty acres or more has been proportionately greater than the small cultivators have although all classes of cultivators have made some gains. The initial advantage gained by the wealthy farmers had a multiplying effect. They had the essential tools and facilities to adopt the high-yielding variety seeds. The increased income netted from this new technology was used to further reinvest in land

and agricultural machinery. The replacement of animal power with tractors and threshers made agriculture still more profitable, permitting cultivators to increase their incomes spectacularly, thereby permitting them to diversify their agricultural operations into other commercial enterprises. Consequently, as Frankel has noted, "farmers with substantial holdings of fifty acres or more experienced a qualitative change in their standard"[53] and a pattern of life that was in sharp variance from the traditional village way of life. This has been reflected in the physical changes noticeable around the villages, such as paved road surfaces, modern brick houses, tractors and high-tension wires. Accordingly, as Day and Singh point out,

> The roads are clogged with traffic and the small market towns are growing into unchecked urban sprawls. There is profusion of shops especially those dealing with agricultural machinery and modern inputs.[54]

With agriculture now a profitable business and the Jat Sikhs' monopoly over the land, which has become a valuable and scarce resource, the Jats have emerged as the new dominant caste in the village hierarchy. They now challenge the former power exercised by the Brahmins. Increasingly, the Jats dispense with the Brahmins as family priests and now call in Sikh priests instead to officiate on important ceremonial occasions. This is in contrast to the situation that prevailed earlier, when Brahmins officiated over all such occasions for Jats and were consequently accorded considerable honor and respect by them. Now, however, the power of the Brahmins in the Punjab has declined, while the Jats' control and ownership of the important land resources have contributed significantly to their overall political and economic power. The strengthened socioeconomic and political power of the Jats has also led to the popularity of Sikhism. Many service-oriented castes, such as carpenters and tailors, are adopting Sikh rituals and customs in order to identify with the powerful landowning Jats. They are also moving away from Hindu customs that were formerly ungrudgingly accepted by all groups, including Sikhs.[55] Thus, significant shifts in the status of various groups have taken place as a result of the recent technological and accompanying economic changes in agriculture.

The Green Revolution and the Politics of Discontent

The Green Revolution precipitated cultural and social changes within the Punjabi society. It was heralded as a technological break-through that would lead Punjab to abundance and material prosperity. For a period of time it did just that. The Green Revolution brought material prosperity and wealth in Punjab. But it also led to increasing inequalities and unmet aspirations that became the breeding ground for ethnic conflict and political violence. Thousands of people were killed due to communal and ethnic violence between Hindus and Sikhs, the two religious groups, that throughout the history of India had coexisted in harmony and peace. The Punjab, the venue of abundance and prosperity for its citizens was plagued by serious problems that brought the State of Punjab into direct confrontation with the central government of India. The conflict in Punjab had ramifications far beyond the boundaries of the State. The conditions that precipitated this conflict and the resulting consequences are described below.

The state of Punjab spearheaded the Green Revolution in India and came to be known as the breadbasket of India's food granary. It was thought of as a land of prosperity and economic progress in post independent India. Social scientists writing about Punjab in the early decade of the 1980's dwelled upon the glory of Punjab's economic achievements due to the green revolution.[56] The Green Revolution of course was a strategy to create abundance of food production by introducing new and modern technological/scientific agricultural practices that would move India from scarcity to abundance. However the economic success experienced in Punjab also had destabilizing affects on the society and polity of Punjab. This is evident in some of the writings on Punjab that focused on its 'crisis.'[57] Two decades after the introduction of the technology of Green Revolution, Punjab was ravaged by violence despite being one of the most prosperous states in India. Ironically it was the creation of prosperity through technological innovation that that led to discontent and disgruntlement in Punjab. This can be traced to two sets of factors. On the one hand it brought Punjab in conflict with the central government of India over resource allocation crucial for agriculture; on the other hand it heightening the sense of both absolute and relative deprivation among classes of people who did not benefit from the 'green revolution'. "The first thrust of the green revolution set off a standard of living and a scale of emulation which became difficult to sustain as the years went by."[58] According to one assessment, the benefits of the Green Revolution were unsustainable in the long run as it introduced a system of food production that was anti-nature and ecologically destructive.[59]

The agricultural processes of cultivation itself had been altered by the Green Revolution. As noted earlier when machines replaced human labor changes occurred in the caste relations among various interdependent groups in the village community engaged in the process of cultivation. The indigenous system of cultivation was labor intensive and depended upon recyclable sources of energy generated from within the society. The system of cultivation associated with the Green Revolution was capital intensive and required inanimate resources such as chemical fertilizers, pesticides and hybrid variety of seeds to increase agricultural productivity. Under this system the farmer became dependent upon external institutions and organizations that controlled these resources required for farming.[60] The chemicals, pesticides etc. were inputs that had to be externally purchased from agencies that procured and managed them. The increasing use of external inputs significantly altered the structure of relations of cultivating farmers internally as well as externally. The former relationship of mutual obligations (although unequal) gave way to dependence on the market making farmers individually vulnerable to the impersonal forces of the market.

After an initial experience of prosperity Punjab farmers were disillusioned. According to one source the rate of returns over costs incurred for wheat cultivation for Punjab farmers dropped from 27.28 percent in 1972–73 to 10.89 percent in 1978–79[61]. The burden of declining rates of return was especially acute among the small farmers with land holdings of five acres or less who constituted the bulk of the farming community. They could not bear the high costs of inputs necessary for this kind of cultivation. There was not enough money to go around after all modern farm expenses were paid for. Many farmers were forced off their land. With the declining returns on agriculture many began to look out for scarce jobs elsewhere[62]. In earlier times all the sons working on the family farm was economic. Under the new system the small family farms had to push its younger sons to go to towns in search of jobs where they had to compete with a wider pool of labor from all parts of India. Many of these young men from villages who were displaced from family farms were also educated. Being educated and unemployed coupled with the growing indebtedness contributed to the feeling of victimization by those viewed as responsible for their impoverishment. Economic discontent soon was translated into political anger with farmers joining in agitation against the central government and its agencies for being treated 'like a colony of the center to feed India'. When their demands were not satisfied a heightened sense of betrayal and frustration resulted. This sense of betrayal was felt along cultural and political lines. The farmers mainly Sikhs saw themselves as discriminated by the central government whose policies were seen as affecting their livelihood in the state. The State of Punjab was embroiled in serious dispute with the central government of India over a variety of issues that were related to sharing of resources vital to agriculture.

As the conflict persisted, the discontent among the disgruntled Sikhs mounted and political agitation became increasingly violent and bloody. Many Sikh militants began to see themselves as a wronged ethnic minority under Hindu domination and started an agitation for a separate Sikh State. When a group of Sikhs militants resorted to political violence to resolve their problems, a volatile situation became even worse. Punjab, the State of agricultural abundance soon became the venue of incessant political violence. Innumerable lives were lost in this political violence as the demand for a separate Sikh nation from a splinter faction of Sikhs in Punjab gained momentum.

Religion and Social Behavior

The economic prosperity brought about by technological breakthroughs in the Punjab has affected the inter-group relationships between religions such as the Hindus and the Sikhs, as noted above. Apart from bringing about relational changes among the groups involved in the agricultural process, it also nurtured Sikh grievance against the Hindus. In this section, the role of religion in the technological and economic development of the Punjab will be examined.

In post independent India Sikhs were among the most prosperous communities raising Punjab's per capita income to the highest in the country[63]. It has been suggested that 'the enterprise of the farmers has lifted Punjab from the quagmire of third world prosperity into second world comfort.' Many people in the Punjab believe that the success of the Green Revolution is partly due to the Sikh religion.[64] They suggest that Sikhs are adventurers in nature and that may underlie their willingness to adopt new agricultural technology. This may be so. But it is difficult to test this relationship in view of the fact that the technology of the green revolution was not available at an earlier period, when the peasantry was burdened with a social structure that provided little support for economic improvement. The Green Revolution occurred at a time when significant structural changes had been initiated. Under the new socioeconomic environment, the religious factor became important in directing the course of economic and political changes occurring in the Punjab. Indeed the rise of Sikh consciousness in Punjab was both due to the political history of the region as well as by the tide and pride of economic success due to the Green revolution. Religion, a significant part of the overall value system of a society, is affected by such social factors as the existing system of social arrangement of people and their relation to valuable resources just as much as it affects the social factors that sustain it.

The Sikh religion has much in common with Hinduism. Both religions are grounded in Vedantic tradition, from which their central values emanate. The concept of bhakti- devotion- is an integral part of Sikhism not unlike that which occurs in Hinduism. It is this idea of personal devotion that is connected to another vital aspect of Sikhism, namely, self-sacrifice. Together, these values constitute the central values of the Sikh religion and have often been invoked to emphasize the interests of the community. It is a relatively young religion on the world scene and is devoid of the hierarchy that marks some of the older world religions.

The central ideas of Sikhism have been used to rally public support on issues that affect the community at large. The organization of Sikh religion has provided a public platform upon which personal issues affecting the collective group are discussed and translated into public concern. The Sikh ceremony is an important medium of translating the fundamental ideas of Sikhism into political and public issues. The one Sikh ceremony, apart from initiation, is the reading of Guru Granth Sahib, its religious scripture. On this occasion, the preacher reads, in part or in whole, the religious text and interprets it to his audience. This occasion is often used to relate the central ideas to current concerns. The preacher, of course, is not free from political influence.[65]

After the independence of India, the Punjab was later divided into two new states, along ethnic lines (on the basis of religion and language), as a result of the political agitation of Sikh leaders who espoused the interests of the Sikh community. The states were the Punjab, which was predominantly Sikh and whose language was Punjabi, and Haryana, which consisted mainly of Hindi-speaking Hindus. When the government of the new Punjab, came into office, it began to implement economic policies that it had argued for. These included land reform and the availability of easy credit, both of which affected the individual farmer. Therefore, the Sikh religion was used to influence and direct the socioeconomic policies of the state government of the Punjab.

The Sikh religion provided a conceptual framework through which individuals could debate organizational and moral priorities that could coordinate individual and community interests.[66] Sikh leaders often used religious ideas

to articulate and mobilize support for policies relating to economic development and social change.[67] This was particularly true in the period following
the independence of India, as well as later, following the separation of the
Punjab from Haryana; even more recently, it has been true in the conflict that
has erupted between the state and the central government that has resulted in
significant political violence with loss of life and property. The recent demand
for Khalistan, an independent nation for the Sikhs, is illustrative of the use of
religion in the service of politico-economic goals.

Conclusion

The case study of agricultural change in the Punjab highlights the importance
of societal features in both inhibiting and promoting technological change.
For centuries, the system of agricultural production continued in its primitive
form, even though political rulers changed. This was due to the social
structures that failed to provide incentives to the farmers to increase
productivity. Measures were subsequently introduced to ameliorate the
position of the agricultural class. But it was only when the obstacles that
inhibited technological innovation were removed that mechanization
occurred. Once the social hurdles were removed, the peasants were quick to
respond to technological changes. Nevertheless, this was not without
consequences. Mechanization brought wealth and prosperity to the farmers,
but it also upset the traditional social ties between groups and contributed to
increasing disparities between rich and poor. Changes in the quality of life
unknown in the rural setting were also introduced as a result of technolgical
changes in agriculture. Significantly, these changes established new patterns of
dominance. The Jat landowning castes henceforth became the most powerful
group in the village hierarchy, and their rituals and customs became popularly
adopted by other castes, contributing to a religious trend toward Sikhism and
away from Hinduism. The Sikh religion itself became instrumental in
directing the course of socioeconomic development in the region. This
resulted from a structural condition of society, namely, the concentration of
large numbers of Sikhs in a contiguous region following the partition of India
into two nation states. The physical proximity of Sikhs, combined with their
socially shared interests, allowed them to use their religion to promulgate their
economic interests. Thus, in the final analysis, technology, society and values
were tied together, each providing an impetus or hurdle for the other,
according to the prevailing conditions of the times.

Notes

1. Barrington Moore, Jr., *Social Origins of Dictatorship and Democracy*, Boston: Beacon Press, 1966, p. 331.
2. Marvin Harris, *Cows, Pigs, Wars and Witches*, New York: Vintage Books, 1947, p. 9.
3. W. K. Stevens, "Farmers of Punjab Are India's Shining Example," *New York Times*, October 7, 1982.
4. *Census of India*, 1971, Series 17, Punjab, p. 4.
5. Neil Charlesworth, *British Rule and the Indian Economy*, London: Macmillan Press, 1982, p. 22.
6. Harris, p. 40.
7. *Encyclopedia Britannica*, Vol. 2, p. 291.
8. Moore, p. 333.
9. Moore, p. 321.
10. H. Calvert, *The Wealth and Welfare of Punjab*, Lahore: Civil & Military Gazette Press, 1922, p. 122.
11. Calvert, p. 122.
12. Charlesworth, p. 16.
13. Charlesworth, p. 19
14. Calvert, p. 122.
15. H. Day & I. Singh, *Economic Development as an Adaptive Process*, New York: Cambridge University Press, 1977, p. 45.
16. Calvert, p. 132.
17. Romesh Dutt, *The Economic History of India*, Vol. II, New York: Augustus M. Kelly Publishers, 1969, p. 84.
18. N. J. Barrier, *The Punjab Alienation of Land Bill of 1900*, Durham, N. C.: Duke University Program in Comparative Studies on Southern Asia, Monograph #2, p. 80.
19. Calvert, p. 124.
20. Darling, *The Punjab Peasant in Prosperity and Debt*, London: Oxford University Press, 1925, pp. 204-205.
21. *The Myth of Population Control*, New York: Monthly Review Press, 1972, p. 52.
22. Calvert, p. 130.
23. Mamdani, p. 58.
24. Alfred Howard, The Agricultural Testament, London: Oxford University Press, 1940.
25. Quoted in Vandana Shiva, The Politics of the Green Revolution: Third World Agriculture, Ecology and Politics, London: Zed Books, 1991, p. 26
26. P. Hershman, *Punjab Kinship and Marriage*, Delhi: Hindustan Publishing Corporation, 1981, p. 21.
27. Barrier, pp. 84-85, and Darling, pp. 208-209.
28. Guy Hunter & Anthony F. Bottrall, *Serving the Small Farmer: Policy Choices in Indian Agriculture*, London: Croom Helm, 1974, p. 149.

29. Amit Bhaduri, "Class Relations and the Pattern of Accumulation in an Agrarian Economy," *Cambridge Journal of Economics*, 1981, p. 150.

30. Hunter & Bottrall, p. 150.

31. Day & Singh, p. 50.

32. Murray I. Leaf, *Song of Hope: The Green Revolution in a Punjabi Village*, New Brunswick, N. J.: Rutgers University Press, 1984, p. 79.

33. Hunter & Bottrall, p. 150.

34. Mamdani, p. 60.

35. Francine R. Frankel, *India's Green Revolution: Economic Gains and Political Costs*, Princeton, N.J.: Princeton University Press, 1972, p. 29.

36. Refer to Mamdani, pp. 66-87, for discussion of family strength and economic resources.

37. Iftikhar Ahmed, "The Green Revolution and Tractorization: Their Mutual Relations and Socio-economic Effects," *International Labor Review*, Vol. 114, No. 1, July-August, 1976, p. 92.

38. Ajit K. Ghosh, "Institutional Structure, Technological Change and Growth in Poor Agrarian Economics: An Analysis with Reference to Bengal and Punjab," *World Development*, Vol. 7, No. 4-5, p. 386.

39. Ghosh, p. 386.

40. Ghosh, p. 393.

41. Ahmed, p. 87.

42. Sir William Roberts & S. B. S. Kartar Singh, *A Text Book of Punjab Agriculture*, Lahore, 1951, p. 145.

43. Roberts & Singh, p. 144.

44. Mamdani, p. 57.

45. Roberts & Singh, pp. 68-89.

46. Leaf, p. 69.

47. Susan George, *Feeding the Few: Corporate Control of Food*, Washington, D.C.: Institute for Policy Studies, 1981, p. 39.

48. George, p. 39.

49. Frankel, p. 40.

50. Keith Griffin, *The Political Economy of Agrarian Change*, London: Macmillan Press, 1974, p. 68.

51. Griffin, p. 77.

52. Frankel, p. 22.

53. Frankel, p. 26.

54. Day & Singh, p. 48.

55. Hershman, p. 23.

56. Jodka, S.S. 'Crisis' of the 1980s and Changing Agenda of 'Punjab Studies', Economic and Political Weekly, February 8, 1997, p.274

57. ibid

58. Gupta, D. The Communalizing of Punjab, 1980–1985, Economic and Political Weekly, Vol. 20, 1985 p. 1189

59. Shiva, p. 11

60. ibid, p. 171

61. Gill, S. S. & Singhal, K. C. 'Punjab Farmer's Agitation-Response to Developmental Crisis in Agriculture' Economic and Political Weekly, vol. 19, 1984, p. 1729

62. Tully, M & Jacob, S. *Amritsar Mrs. Gandhi's Last Battle*, London: Jonathan Cape, 1985, p. 49

63. ibid, p. 36

64. Leaf, p. 180.

65. Leaf, p. 188.

66. Leaf, p. 180.

67. Murray I. Leaf, *Information and Behavior in a Sikh Village*, Los Angeles: University of California Press, 1972, p. 162.

CHAPTER 6

The Industrial Revolution:
The Case of Lowell, Massachusetts

The Setting

The farming community of East Chelmsford in the early nineteenth century became America's first planned industrial city. It was later renamed Lowell after Francis Cabot Lowell, an international trader who, having proved successful in many business ventures, sought new challenges. During a trip to England, Lowell, fascinated by power-driven machinery used in the textile industry there, memorized the design of the English power loom. He returned to Boston with a plan for establishing such an industry in the United States.

Lowell and his partners—a small group of Boston merchants who became known as the Boston Associates—first set up a small textile mill in Waltham, Massachusetts, on the Charles River.[1] Finding the water power from the Charles River too sluggish for a big operation, they consequently searched for a better source. The potential power required was located thirty miles north of Boston in the village of East Chelmsford, which was situated on the Merrimack River at the point of the Pawtucket Falls. The water from the Pawtucket Falls would be harnessed to feed the power canals that were to be connected to a series of textile mills. The energy generated would be sufficient to drive the machinery. East Chelmsford, with its natural drop of thirty feet at the falls, and its location at the confluence of the Merrimack and Concord rivers, was the ideal site to provide abundant water power and cheap river transportation.[2] Land was cheap and available. Young women from nearby farms would be available to tend the power-driven looms in the textile mills. Life on the farms would remain undisturbed. Fathers and sons would continue in their traditional roles.

On November 1821, this small group of Boston merchants launched what became known as the Lowell experiment. They pooled the capital they had accumulated from shipping and trading activities and utilized Francis Cabot Lowell's improvements on the English machinery. They also put into practice their ideas of cooperative paternalism, adopted from the earlier Waltham experiment. This set the stage for the industrial revolution in America and the process that would transform rural East Chelmsford into an industrial center of redbrick factories and boardinghouses. Indeed, East Chelmsford became the prototype for the economic reorganization of American life.[3]

For more than a century before the cotton factories appeared along the Merrimack River, the inhabitants of the area used the water power of the Merrimack and Concord rivers in a limited way for cutting timber, grinding grains and driving home weaving. They lived as Thomas Jefferson extolled people to live—a slow-paced existence tuned to the seasons, with every household an independent unit.[4] On the farms surrounding East Chelmsford, none foresaw the consequences of the swift and irrevocable transformation that would soon occur. Alexander Hamilton's proposal that the United States become a prosperous nation through the development of large-scale manufacturing and trade first was tested in this quiet village.

By 1826, Lowell had attracted enough people—approximately two thousand—to become a separate town. By 1830, a considerable industrial town arose where only ten years earlier there had been farms and a few small mills. In 1836, Lowell was incorporated as a city with a population of almost twenty thousand.

The first company formed by the Boston entrepreneurs was the Merrimack Manufacturing Company. After it began operation (it took only two years) and turned out its first finished cloth in 1823, the need for more mills, canals and machinery was apparent. The immediate success of the Merrimack Company and the expansionist vision of the Boston Associates led to the founding of a series of new firms in rapid succession.[5] A new form of corporate system was initiated.

Each of the large plants was designed to accommodate the revolutionary technology. The pattern developed required a large plant where the total process of manufacturing from raw materials (cotton) to finished product (cloth) was undertaken. There was the combining of large capitalization with professional management, and for the first time, closely supervised labor was contained within the factory walls for all the manufacturing stages. Each of

Lowell's mill complexes housed a complete production unit. Wherever possible, machines replaced hand labor, thus greatly increasing productivity and lowering costs.

The outward appearance of the mills as well as their interior design and organization reflected uniformity, order and regularity, essential elements for the manufacturing process. The basement housed the waterwheel, placed below ground level to maximize the power generated and to protect water in the millrace from freezing temperatures. Successive stories housed the carding, spinning, weaving and dressing steps, each operation occupying a single room on a floor. An elevator connecting the different floors moved materials from one step in the production process to the next.[6]

The organization of work was aimed at speeding the flow of materials. The carding room's location on the ground floor was determined primarily by the heavy weight of the machinery and its proximity to the picking house, where machinery opened and cleaned the baled cotton. The initial carding and drawing operations transformed the loose cotton into a coarse roving that spinning frames on the second floor twisted into yarn. On the upper floors of the mill, warping and dressing machines prepared yarn for the weaving process and power looms produced the finished cloth. A single cloth room serviced all the mills of a company, and there operatives measured, folded and batched the fabric for subsequent bleaching, dying or printing and eventual shipment to selling agencies in Boston.[7]

Combining utopian social concepts with keen business practices, the Boston Associates desired to make the Lowell system a model for industrialization in the country as a whole. Rapid industrialization occurred, and by 1848 there were ten flourishing mills and eight power canals. In the early years, the mill owners took advantage of an untapped supply of cheap labor—young single women from the surrounding farms. The technology of the power loom required only semiskilled workers. Highly skilled male workers were not essential to the process.

The paternalistic policies of the mills were designed to make profits for the mill owners and to provide a wholesome atmosphere for the work force. Industrialization was to prosper in a rustic setting. The founders of the company wished to avoid the squalor and degraded working conditions that prevailed in English towns such as Manchester. They desired to create a model community uniting the cherished traditional values of agrarianism and the new values of manufacturing. An image of America that romanticized the virtues of farm families—self-sufficient and independent—loomed large. There was admiration for British technology and productivity but not for its social consequences. The aim in the United States and Lowell in particular was to avoid these pitfalls. Thus, the Lowell conception embodied a planned industrial town.[8] Glossed over was the inevitable conflict posed for capitalists, who, while promoting technology for profit, were supposed to protect the common good.

Introduction of the Factory System: The Yankee Women

The employment of New England farm women was not expected to alter the fabric of the family structure. Those in charge assumed that the family economy would remain strong. Making cloth had always been women's work, and the skills of spinning and weaving were household arts. Therefore, even though production moved to the mills, women would still perform traditional tasks. A shift in place of work, the provision for wages and the use of machine technology were not perceived as bringing changes in human relationships. There was little understanding that new productive forces would foster new social roles.

New England farms had never prospered. There was always the need for additional income. The young women were not indispensable to the farm operation. The stimulus for women to work outside the home was provided by the values of the nuclear family unit, with its emphasis on autonomy and individualism. The Jeffersonian ideal of the independent yeoman farmer personifying the virtues of self-sufficiency was left intact. Manufacturing would be introduced without disturbing the nation's agricultural base. Farmers would still provide the underpinnings for a stable democracy.[9]

The women were expected to make a brief commitment (a few years at most) to mill work, then return home, marry and raise families. Life would remain orderly both in the mills and in the countryside. The women would live in company boardinghouses adjacent to the mills. At all times, they would be under the disciplined control of the mill supervisors. On Sunday, church attendance was required and money was deducted from the women's wages for church dues. These practices served to reinforce the close links between traditional Protestant religious values and the development of industrialization. The mill agents and overseers were members, teachers and often deacons

The Mill Girl by Winslow Homer
Courtesy: Merrimack Valley Textile Museum

in the same churches the young women attended. Hence, mills and churches were intertwined.As a moral factory town, Lowell would provide a crucial link in making manufacturing acceptable. The docile women workers were not expected to promulgate any form of labor unrest.[10] They would not view themselves as a proletariat. The Boston Associates planned to protect their economic interests but at the same time totally disassociate themselves from the British model of industrialization, where demoralized laborers might pose a threat by demanding improved working conditions or disintegrate into a rabble. Their motto was "One who made goods could be a fit companion for one who tilled the soil." Through the medium of the Lowell mill women, Francis Cabot Lowell and his cohorts intended to prove that industrial technology and republican values could coexist.[11] They did not anticipate that a city organized for manufacturing might prove incompatible with maintaining idealized pastoral surroundings.

Their grand scheme did not provide a place for immigrant workers. Thus, the Irish laborers who came to Lowell to dig the canals were ignored. They lived on the edge of the city. Little thought was given to the developing political and social organization of the community. There were no plans for involving the general population in decision making. The women workers did not participate in civic activities or town affairs. Increasingly, an urban identity separate from and often in opposition to the manufacturing enterprise developed.

Within a decade of Lowell's founding, the attempt of the mill owners to maintain a paternalistic system began to unravel. Change was impending. The social and economic foundation of the city was becoming diversified. The middle landscape, the judicious blending of city and countryside, was disappearing. European immigrants began to replace the Yankee women in the workplace. The cumulative impact of more demanding work and declining earnings led some women to abandon the mills and discourage new recruits. As the new immigrants, most often in dire circumstances, arrived in Boston or came south from Quebec, mill agents had an abundant supply of workers to fill the empty places. In 1845, potato blight destroyed great portions of Ireland's food base, resulting in a mass migration to America.

Once the women left the farms to work, family relationships were never again the same. Overall, the women sent their wages home to help maintain family homesteads or to provide funds for brothers to go to college. Undoubtedly, some came to gain a measure of economic independence, which, in turn, enabled many to chart new paths. The separation of place of work from family had important consequences. Acceptance by the peer group of other workers often superseded loyalty to family. The boarding housekeeper became a surrogate parent and was often called "Mother." The constant social contacts at work and at rest led to a solidarity of interest among the women. This network of friendship provided a protection against loneliness.

Many women did not come as isolated individuals but as members of a broader kinship structure. There were many sister pairs, and many operatives had cousins employed as well.[12] The existence of these bonds eased the shock of adjustment to factory work and urban living. Still, some were caught up in the shifting of values, the results of structural changes taking place in the transition from preindustrial to industrial conditions. The values associated with industrialization equated time and money. Neither should be wasted. The rhythms of life were altered—physical movements were attuned to machines. The women worked at an entirely different pace from that associated with farm activities. Farm work had been task-oriented, intense labor followed by periods of rest.

Farmers' daughters were thrust into work routines that demanded at all times subservience to machines. In the early years, the women were not required to produce as fast as in later periods.

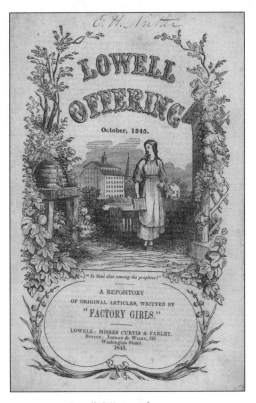

Lowell Offering title page
Courtesy: Lowell Historical Society

Merrimack Manufacturing Company employees, circa 1880
Courtesy: Lowell Historical Society

In the initial industrialization phases, they tended fewer looms and interspersed their routine with reading and caring for the many plants that decorated the mill windows. Later, when Lowell's dominance was challenged by outside competition, the emphasis on profits prevailed. This required greater productivity and, as the pace of work quickened, the women tended more looms. Money established its own scale of values.

Several of the essays that appeared in the *Lowell Offering*, a literary publication written and edited by the mill women, provided clues to changes in behavior required for adaptation to machines. The noise of the machinery was deafening; many women complained of a ringing in their ears even when at rest. Physical confinement also posed problems that resulted in unnecessary hardship on women who had grown up surrounded by the sights and sounds of nature. In a brief time span, the new technology required unanticipated changes in behavior patterns.

There were also health problems created by the new machinery. Many operatives developed lung problems from breathing the cotton waste that permeated the workrooms. Others suffered nervous disorders brought on by confinement in close quarters and monotonous work. Both the social and physical organization of the mill system imposed restrictions. One mill worker poignantly expressed her sentiments in trying to cope with such conditions when she wrote in the *Lowell Offering*, "I am going home where I will not be obligated to rise so early in the morning nor be dragged about by the ringing of a bell, nor confined in a close noisy room morning till night."[13]

Undoubtedly, the Boston Associates maintained some interest in the physical and mental well-being of their employees; however, it did not extend to an understanding of the psychological impact on workers of adjusting to machines. The founders provided churches, schools and literary society lectures; however, the women worked six-day weeks, of twelve- to fourteen-hour days, thus leaving little time to engage in other activities. The mill owners, served by their mill agents, maintained a narrow vision of the overall economic and political growth of the city. Their goal was the development of a successful business enterprise. The city was viewed as an extension of the mill.

In the initial phase of industrialization, stretches of landscaped grounds surrounded the mills; the countryside, with open views of the river, was largely left intact. Later, as the land became more valuable, the boardinghouses and mills were strung together, and additional floors were added to mill structures. In time, all open spaces were eliminated. The idealized imagery of Lowell, symbolized by the cover drawing of the *Lowell Offering*, which depicted school, church and factory coexisting in perfect harmony, became more a myth than a reality. Capitalism and environmental exploitation went hand in hand.

Following the visit of President Andrew Jackson in 1823, a parade of distinguished men and women made pilgrimages to the celebrated town. Charles Dickens was among the many famous writers who came to observe this industrial utopia.[14] Most visitors were awed by the technological splendor. They praised the moral character and fresh, clean appearance of the women workers. For some, even the machinery represented perfection and beauty. They wrote passionately about the rewards of disciplined factory labor. A local physician, Elisha Bartlett, undertook a study of Lowell workers, purporting to find that the regularity of their work habits contributed to good health. He reported that active labor and light activities aided the avoidance of the evils arising from the extremes of indolence and overexertion. He pointed out that the bland routines provided a positive life-shaping force.[15]

A few writers were more critical. Their appraisals concluded that the great majority of the women were sacrificing their health and spirit to the company while not benefiting themselves at all. The corporations were viewed as not protecting women but instead spawning a new female caste. Attention was drawn to the great social gulf existing between the daughters of the mill owners and agents and the farmers' daughters who labored in the mills. The social critic and reformer, Orestes Brownson, claimed that this gulf was even greater than between nobleman and tenant farmer.[16]

Herman Melville, in his short story "The Tartarus of Maids," delivered a devastating attack on factory routine:

> Not a syllable was breathed. Nothing was heard but the low, steady, overruling hum of the iron animals. The human voice was banished from the spot. Machinery—that vaunted slave of humanity—here stood menially served by human beings, who served mutely and cringingly as the slave serves the Sultan. The girls did not so much seem accessory wheels to the general machinery as mere cogs to the wheels.[17]

Around the mid-1840s, life in the mills grew increasingly less attractive to the young women. Initially, the mills experienced tremendous growth as their technological innovations led to increased productivity. Profits and dividends rose to a peak in 1845, but then dipped substantially. Lowell and other New England textile towns faced increased competition. As a result, Lowell manufacturers sought to increase output in relation to labor costs. They constructed larger mills now powered by steam instead of water. The average size of the new mill, about 6,000 spindles in 1835, grew to 18,000 in 1847 and to 50,000 in 1883. Additional looms were added, and this increased the noise level. The number of operatives employed did not keep pace with the increased mill capacity. New technology made it possible for workers to tend more machines. Overseers, rewarded by bonuses, drove their workers harder to stimulate production. The comparatively slow pace of the 1820s gave way to more intense factory discipline, a lowered rate of pay and a work speedup. Tighter supervision accompanied the faster work pace. The workday was lengthened between 1829 and 1841. Amenities such as conversing on the job and tending potted plants in the windows came under attack.

The women expressed their discontent over the changed work conditions by seeking to obtain a ten-hour day. By 1844, reducing the workweek became a key issue. Led by Sarah Bagley, an employee of the Hamilton Company, a group of operatives took over a labor newspaper, The Voice of Industry. It became the counterpart of the dying Lowell Offering. Bagley submitted two thousand signatures in support of the ten-hour work day to the Massachusetts Legislature.[18] This act took great courage. Those the mills deemed agitators and troublemakers could be blacklisted and thus never again allowed to work in the mills.

The movement for a shorter workday failed and a turnover of workers began. Women who had accommodated the demand of the manufacturing process now began to rebel against it. They valued their independent spirit as daughters of free men. The principles enunciated in the American War of Independence still captivated them. Wage cuts and longer work hours struck the women as an attack on their belief in justice. These policies were viewed by the women not merely as economic acts but as attempts to enslave them.

Protestant religious beliefs that stressed responsibility and accountability for one's actions also played a part in their disenchantment. The Puritan ethic of the women demanded diligence in a calling useful to both the individual and society. They felt, however, that economic activity could not stand apart from the moral considerations of their treatment in the workplace and their lack of control in the work process.

By and large, the women did not return to the declining family farms; instead, they pursued other wage occupations. Some who did go home did not like the isolation; they did not marry farmers as had been expected. Still others got caught up in the fever of the movement westward. The positions left vacant by the women were filled by Irish immigrants who made no demands on the factory system, were readily available and also were willing to work for less money.[19] This immigrant work force was not originally envisioned. The ranks of Irish laborers who came to dig the canals had swelled as a result of the devastating famines in Ireland. They were soon joined by French Canadians migrating down from the provinces and later still by European peasants escaping the intolerable conditions of their mother countries.[20]

With the employment of the Irish, who had lived on the very fringe of society, a new niche was established in the mill hierarchy. At the top there were, the company agents, next the overseers, then the mill women, and at the bottom the Irish day laborers. The town of Lowell itself became highly stratified.[21] A middle class largely composed of merchants and professionals provided services. The capitalist class—the mill owners—formed an absentee aristocracy. There were no bonds formed between them and the new proletariat work force.

The Irish and the other immigrants who followed were destined to become a permanent, unskilled labor force. The opportunities and protections provided to the mill women were discarded. Few rewards or amenities were provided immigrant workers. They lived in substandard housing. Their children received little or no schooling. Social and economic circumstances influenced the decision of immigrant youths to work in the mills instead of attending school. Indeed, the use of child labor was an established practice. It was accepted without question that the services of the immigrants were entirely for furthering industrial growth.[22]

The technological development of a new power provided by the harnessing of water and, later, steam, coupled with a new organization of production, set in motion a transformation. A watershed was crossed with the transition from a family economy to a labor force comprised of unmarried women workers and then immigrants. In time, a complete revolution in domestic and social life occurred. After industrialization, even the family farms were tied to a money economy. The large manufacturing establishments, with their strict division of labor and vertical structure, established a new exchange system where the manufacturers asked nothing of the worker but his labor. The laborer, in turn, expected nothing but his wages. With the shift to immigrant workers, ties between owners and workers were more and more defined within narrow limits of supply and demand.

The machine was more than labor-saving; it became life-shaping as well. In the past, the laborer manipulated his tools. In contrast, the machine is characterized by its automated action. The skill level of the worker becomes less and less important. The machine has an imperative of its own, and the worker develops a passive relationship to a mechanical production system.

Immigrant Work Force

The Irish began to enter the mills at a time when company control of its traditional labor pool was increasingly unstable. Some of the New England women left the mills because they would not compete with the cheaper foreign labor. Only seven percent of the operatives were Irish in 1845, but by the early 1850s their proportion was estimated at one-half, and it grew higher year by year. With the arrival of the Irish, Lowell developed a permanent factory population. The life of the immigrant family was completely dominated by the industrial machine. Men, women and children labored in the mills. Most often it was necessary for entire families to work in order to survive. Accordingly, in some sense, the family as a work unit was reintroduced.

The new workers, with internalized values associated with their peasant backgrounds, had a difficult adjustment to make. Fathers lost authority as children, wives and unmarried daughters became wage earners. Often it was easier for women and children than for men to gain employment in the unskilled jobs. Machine improvements made jobs less complex. Irish children were employed directly in the production process. Most worked in the spinning rooms in the traditional child occupation of bobbin boy or girl. The Irish faced prejudice and discrimination; the best jobs were reserved for the native Protestant population. This practice continued into the twentieth century. Mill management and other administrative positions still remained in the hands of the Protestant establishment.

With the introduction of the immigrant workers, Lowell was on its way to becoming a factory town. The class system became more stratified. The Irish and each of the ethnic groups that followed were treated as appendages to the machines. As the machine process became more automated, jobs increasingly became unskilled. The Yankee women had maintained some status at work, but as the immigrant workers took over, factory work became degraded.

There were inadequate medical services; tuberculosis and typhoid were rampant and, diet was poor.[23] Those at the top of the class hierarchy showed little concern for the well-being of the workers. As Cardinal O'Connell stated, in writing of his youth in Lowell,

> These newcomers, first entirely Irish, later French Canadian Catholics from the provinces were treated precisely as if they were part of the machinery which ground out the millions being produced for the rich managers and mill owners who spent the money not in Lowell, but in New York, Boston, Paris and London.[24]

He further wrote that the workers got all the hard work and almost nothing else, certainly not compassion, pity or understanding of the almost insupportable conditions of labor at that time.[25]

Over time, the new immigrant workers did begin to press for better conditions. There were major strikes in 1875, 1903, 1912 and 1934.[26] Unlike the Yankee women, who gave the highest priority to shorter hours, the immigrants' main concern was for better pay. The immigrants were deeply religious and accepting of the conditions they faced. Although Catholic priests were sometimes sympathetic to Catholic workers in their opposition to Protestant mill owners, overall religious values reinforced the status quo. Militancy and social unrest were not condoned.

The church did not effectively speak out against the obvious abuses of the industrialists nor did it aid a drive for unions. Kirk Boott, the first resident agent of the Merrimack Company and a staunch Episcopalian, went out of his way to help the Catholic church establish a parish in Lowell. He hoped that the discipline and authority of the church would keep the predominantly Catholic workers from making demands on the system.[27]

An organized labor movement was slow in starting and suffered many defeats. It was retarded in part by the constant influx of new groups always desperate for work. The immigrants provided an almost inexhaustible supply of cheap, tractable labor. The acceptance of unskilled jobs in the mills represented a last stand against pauperism. The immigrants took it for granted that all members of the family would work, as soon as possible, and offered little opposition to child labor. The job of tending machines allowed little opportunity for the development of skills or advancement up the industrial hierarchy. The repetitive job routines did not allow for work teams that might foster cooperation. The lack of personal freedom and initiative on the job was not conducive to union organization.

Ethnic divisions often interfered with worker solidarity. The nationality groups were separated each from the other by barriers of language and culture. Each lived as a closed community isolated from the rest. Close ties and loyalty to others did not extend beyond family, church and neighborhood. Each Catholic ethnic group maintained its own national church. By contrast, in England, the proletariat population remained stable and homogenous and therefore was more easily organized. As early as 1847, the ten-hour day became law in England. It was 1874 before the ten-hour day became law in Massachusetts.

The expected passivity of the immigrant workers was somewhat shaken by a strike of female Irish throstle spinners in 1859. Also, in 1875, the male mule spinners, a highly skilled group, struck for higher wages. The machinery they used at the time required physical strength. The mule spinners had to drive the mule along its track, thus manually guiding the yarn as it was spun. Management, however, undermined their efforts by employing a new technology, ring spinning, which required less skill. Unlike the mule, the ring-spinning machine was stationary. The development of the automatic loom—if a thread broke, the machine stopped—enabled one person to tend several machines. This, in turn, led to speedups and stretch-outs. New technological developments undercut skills and wages. Labor unrest often followed these changes.

In Lowell, the absentee owners were removed from any personal accountability. Their brand of industrial capitalism, which focused only on how much cotton cloth was produced and at what cost, was not challenged. The company founders were not interested in developing community spirit. Their objective was the successful operation of a business enterprise. The mills operated as one big monopolistic unit. The values of the market place prevailed. There was price-fixing and restricted sale of stock. A few men appear time and again among the directors of these various companies. These interlocking directorates assured that no single firm would place its own interests above those of the larger group of firms.[28] Members of intermarrying families dominated decision making.[29]

After a peak in 1845, profits and dividends sagged. There was a weak upturn in 1851-53 until the onset of the Civil War. During the war, the cotton industry in Lowell all but closed down. Irish men went into the Union Army and cotton was unavailable from the South. The Civil War years were a dividing wedge for Lowell's industrial fate. The successes of the prewar decades were never again realized. Increasingly, profits were made at the expense of the mill operatives.

After 1845, Lowell's golden age ended. By the early 1850s, the Irish were well established in the work force. Throughout the remainder of the nineteenth century, industrial decline accelerated. By 1900, the native population had given up political control to the Irish. During the 1920s and 1930s, the great textile mills closed. By 1940, only three of the original companies remained. Depressions, unemployment and economic dislocations plagued the city.[30] In the 1970s, a modest revival began with the introduction of a new technology: a computer industry was established in the city.

Conclusion

The Lowell system of industrialization, which featured large capitalization, corporate ownership, management by salaried agents and a predominantly female labor force lodged in company boardinghouses, spread throughout New England in the early nineteenth century. A technology based upon the power loom and the employment of well-educated local women as semiskilled workers achieved high productivity and large profits. The plentiful supply of women who would remain for only a brief tenure was expected to provide a golden age for Lowell. The women would be under the paternalistic guidance of mill agents and would not be reduced to a permanent proletariat.

By 1830, change became imminent. The idealized factory in the bucolic setting was disappearing and an urban identity was taking hold. The manufacturing enterprises that presented their goals as synonymous with city development did not exist in fact. Shaped by judgments made in the marketplace, the city increasingly was swamped by uncontrolled and chaotic growth. The well-built brick mills set in an orderly landscape were sacrificed to profits. By 1850, the concept of Lowell as a prototype for what manufacturing should be in America was largely abandoned. The republican values of the women affected the technological process. They left the mills rather than work under intolerable conditions. They resisted the demands of factory discipline. The impact of working for cash wages and the destruction of the family economy meant their lives would never again be the same.

As the immigrant workers replaced the Yankee women, working conditions deteriorated and wages were lowered. These workers became a new proletariat. There was no longer any pretense of combining agrarianism and industrialization. The factory system led to rigid stratification. The workers were locked into jobs at the bottom of the skill hierarchy. Over time, unionization did bring some modest improvements in working conditions, and some workers, through education, did escape the mills.

Overall, the fate of the workers was intertwined with the declining profits of the mills. The utopian dream disintegrated into the reality of poverty, slums and smoking factories. When in the 1930s and 1940s most mills ceased operation due to fewer markets and a base of cheaper operation in the South, the city was all but shut down. Lowell had been founded as an economic institution. Its dependence on only one form of technology and the growth of monopoly capitalism hastened its decline. The technology and values associated with industrialization took precedence over most former values. Profound changes in life patterns were ushered in with the shift to work for wages in a workplace separated from family life.

The preindustrial era had required task-oriented work operations interlaced with frequent periods of rest. The power-driven machines required constant attention and adaptation on the part of the workers. Industrialization eroded the social foundations of the rural countryside. The independent

Yankee farm women were replaced by an interdependent urban work force that operated within the framework of a highly stratified society. The industrial process as it unfolded in Lowell dramatically illustrates that technological innovations inevitably bring changes in social structure, culture and values.

Notes

1. Frances W. Gregory, *Nathan Appleton: Merchant and Entrepreneur, 1779-1861*, Charlottesville: University Press of Virginia, 1975, p. 158.

2. Gregory, pp. 173-193.

3. Thomas Bender, *Toward an Urban Vision: Ideas and Institutions in Nineteenth Century America*, Lexington: The University Press of Kentucky, 1975, pp. 29-51.

4. John A. Goodwin, "Villages at Wamesit Neck," Chapter 5 in *Cotton Was King, A History of Lowell, Massachusetts*, Arthur L. Eno, Jr., ed., Somersworth, N. H.: New Hampshire Publishing Company, 1976, pp. 57-79.

5. Thomas Dublin, *Women at Work: The Transformation of Work and Community in Lowell, Massachusetts, 1826-1860*, New York: Columbia University Press, 1979, p. 206

6. Dublin, pp. 61-62.

7. Dublin, pp. 61-62.

8. Bender, pp. 97-128.

9. John F. Kasson, *Civilizing the Machine: Technology and Republican Values in America, 1776-1900*, New York: Grossman, 1976, pp. 3-51.

10. Herbert G. Gutman, *Work, Culture and Society in Industrializing America*, New York: Vintage Books, 1976, pp. 25-30.

11. Kasson, pp. 53-106.

12. Dublin, p. 42.

13. *The Lowell Offering*, prepared by females employed in the mills, Lowell, 1840-1845, New York: Greenwood, 1970.

14. Charles Dickens, *American Notes*, New York: Harper and Brothers, 1842.

15. Elisha Bartlett, *A Vindication of the Character and Conditions of the Females in the Lowell Mills*, Leonard Huntress, printer, 1841, pp. 6-13. This report is also discussed in Bender, pp. 64-65.

16. Orestes A. Brownson, "The Laboring Classes," *Boston Quarterly Review*, Vol. 3, July and October 1840, pp. 364, 472-73.

17. Herman Melville, "The Tartarus of Maids," *Great Short Works of Herman Melville*, New York: Harper & Row, Perennial Classic Edition, 1969, pp. 215-216.

18. Philip S. Foner, ed., "Sarah Bagley," Part 4 of *The Factory Girls*, Chicago: University of Illinois Press, 1977, pp. 159-177.

19. See discussion of the early Irish community in George F. O'Dwyer, *The Irish Catholic Genesis of Lowell*, revised edition, Lowell, Mass.: Lowell Museum Corporation, 1981.

20. See Charles Cowley, *The Foreign Colonies of Lowell*, Lowell, Mass.: Old Resident Historical Association, 1881. Also see Shirley Kolack, "Lowell, An Immigrant City: The Old and the New," in Roy Bryce-Laporte, ed., *Sourcebook on the New Immigration*, New Brunswick, N. J.: Transaction Books, 1980, pp. 339-345.

21. Harriet H. Robinson, *Loom and Spindle, or Life Among the Early Mill Girls*, Hawaii: Press Pacifica, 1976, pp. 8-9.

22. Peter Blewett, "The New People: An Introduction to the Ethnic History of Lowell," in *Cotton Was King*, Arthur L. Eno, Jr., ed., pp. 190-217.

23. George Kenngott, *The Record of a City*, New York: Macmillan. 1912, pp. 28-44.

24. Cardinal William O'Connell, *Reflections of Seventy Years*, Boston: Houghton Mifflin Company, 1934, p.12.

25. O'Connell, p. 12.

26. See Mary Blewett, ed., *Surviving Hard Times, The Working People of Lowell*, Lowell, Mass.: Lowell Museum, 1982. Also see for treatment of strikes in the Merrimack Valley area, Donald Cole, *Immigrant City*, Chapel Hill: University of North Carolina Press, 1963, pp. 94-95.

27. Joseph W. Lipchitz, "The Golden Age," Chapter 7 in *Cotton Was King*, Arthur L. Eno, Jr., ed., pp. 98-99.

28. Dublin, p. 10.

29. Gregory, pp. 17-18.

30. Fedelia O'Brown, "The Decline and Fall: The End of the Dream," in *Cotton Was King*, Arthur L. Eno, Jr., ed., pp. 141-158.

The Computer Revolution:
The Case of The State of Israel

The Computer Revolution has ushered in the Information Revolution. With the introduction of the computer—a radical transformation has occurred which rivals the effects of the Industrial Revolution. Just as the Industrial Revolution brought changes in all aspects of society, the computer has created wired societies, where people all over the world can instantaneously be in contact with each other.

Thus, the computer provides the connecting links in the often-fragmented world in which we live. Extending from the workplace to the home, we are reliant on the computer for communication and as a source of information.

The stresses of the Industrial Revolution were largely physical with workers moving from their homes. Where there had once been close extended family and village ties, there were now factories filled with strangers in crowded polluted conditions. Work tended to be physically tiring, repetitive and confined to restricted places and times. The Computer Revolution has by contrast changed the emphasis from physical work to mental work.

While people in agricultural societies worked the land around their homes to the movements of the sun, industrialization created the time clock and separate workplace. Wired technology by contrast allows workers to live anywhere. Therefore, many people choose to leave company headquarters and work in their homes. The nine to five workday is disappearing.

The majority of workers in advanced nations are now employed in the service sector. The range of services is extremely broad. A high proportion of service jobs involve various aspects of creating, collecting, analyzing, interpreting and the dissemination of information—all of which require computer literacy.

Computer technology provides the technical links to viewing the globalization process. The capacity of computer instruments to store, memorize and transmit billions of pieces of information in seconds makes the accumulation of information a powerful tool of social control. Traditional values that were lost in the era of industrial manufacturing, but were prevalent in the days of the cottage industry, when the entire family worked from the home, may re-emerge in the new information era.

Computers are especially useful for storing and retrieving information in databases. This capacity of computers is an invaluable tool for researchers as well as providing the general population with quick access to knowledge.

The use of computers provides benefits to the functioning of societies at many different levels. Personal computers may be used for networking through the use of email, electronic homework, education, teleconferencing, Internet shopping, and an abundance of other informational services such as word processing. Computers of all types are used for telephone service, automated banking tellers, military applications, and artificial intelligence. In the near future, nearly everyone will be carrying some sort of communication handset, be it a cell phone, a palm pilot, or a laptop computer.

Corporations are dominant forces in shaping national economies world-wide. This is done through the technological advances that computers promote. For example, greater efficiency, productivity, command, and control. Multinational corporations can have immediate and efficient communications with company branches all over the world. The primary segments for use of computers are the government, corporations, the military, and many small businesses.

New technology inevitably results in changing value systems. The two are inextricably linked together. Personal home computers that allow global networking, without hierarchical structures or centralized administration or authority, are portent equalizers in society. An example of the changes computers bring is the printing trade; where a significant number of jobs have been lost because of computerized typesetting systems. In some manufacturing institutions, workers are replaced by robots, and mid-level management is replaced by computer assisted designs.

However, not everyone views these societal modifications in a positive light. In his book, the Lexus and the Olive Tree, Tom Friedman describes the effects of the computer revolution on society as a part of globalization, or the increasingly integrated system that creates the possibility of a single global economic structure. The Internet allows people to connect to one another immediately, wherever they are. Tom Friedman points out that, "A man can be hiking Mount Washington in New Hampshire, and receive an email via his cell phone from a friend in Shanghai".[1] During the 19th century, impersonal letters acted as the main form of communication; these letters took a long duration of time to arrive. With the advent of the telephone, the time issue was resolved. However long distance phone calls tend to be expensive, and it is difficult to be spontaneous in such situations. Computers have solved all of these problems by providing an instant transfer of any type of information, along with the possibility of face-to-face communication. Yet, being connected at all times through cellular or Internet links can act as a leash to family, coworkers, and most controversially, the government. This can take away from the ability to become a complete individual, as one is constantly available for others' criticism.

The Internet is the cause for a great deal of change in society. It is indicated that technological advances may "enhance government power to survey, regulate and enforce"[2]. On the other hand, technology will allow the individual more freedoms, as well. Many question the regulations of technology in addition to how it will affect the thinking of individuals and the function of the government. According to Joseph Weizenbaum, a computer science professor at MIT said that it was duty of scientists to reveal the negative aspects of their research. Often, computer use initially empowers humans but may eventually render them powerless. A lot can be lost due to computer use. For example, "The lost forms of knowledge are those that depend on the face-to-face human communication, that emphasize tacit awareness, and which are part of larger sensory interactions"[3]. Another disadvantage is that the high technology develops a "work free society" where computers add to the growing income disparity, and replace jobs. High-technology jobs are not nearly enough to replace the ones that have been lost. Many times malfunctions and miscalculations can lead to catastrophic events[4]

Some examples of possible societal alterations include the destruction of traditional business models and Internet voting[5]. According to Marc Grossman, the United States Under Secretary for Political Affairs, these types of transformation necessitate learning to use the revolution to one's advantage[6]. Therefore, the government has greatly participated during the transition into the information revolution. Some of the modifications provided by the Internet are making international interviews available in full text; posting newly released reports in various translations; and increasing the reach into other countries[7]. In fact, the top three of Tom Friedman's *Eight Habits of*

Highly Effective Countries are wiring a country, the speed of the country, and whether or not the country is harvesting its knowledge[8].

These are some of the more generalized effects of the computer revolution. However, this informational transformation has a tremendous effect on individuals, as well. For example, medical doctors based in urban centers can be linked by computer systems to rural regions. Computer component manufacture and access to the technical knowledge of how to build computer systems could affect developing societies occupationally and technically. More specifically, Israel is reshaping its economy to incorporate new information technology and could well reemerge from its problems stronger and in competition with Japan and the US. In fact, the United States is seeking to utilize Israel's highly educated, innovative, professional workforce. Israeli engineers and layout designers can develop silicon chips, software and printed circuit boards. But, Israel is not only popular because of its educated workforce but also because professionals are paid lower salaries than in the US, and the government subsidizes the development of high technology. Unfortunately, there are fears that the ongoing political tension could dampen interest in high-tech industry. For the most part, Israel start-ups were export-oriented, which avoids political conflicts.

Additionally, the recent presence of women in the high-tech workplace has consequences. According to a study by the Israel Women's network, while women without children demonstrate contentment with their employment experience, while women who are mothers have little personal time, and encounter challenges such as locating proper childcare. Nevertheless, women have yet to be equally represented in the high-tech workplace. In fact, at Waukesha County Technical College, less than eight percent of high-tech classes are comprised of women.

Case Study for Israel

Israel's economic future was once based on agricultural exports. Today, innovations in science and technology are creating limitless potentials. In the absence of natural resources-a barren land-the key to Israel's economic viability is development of its most important asset-brain power. High-tech has become the messianic hope of Israel, a silicon savior.

Israel has some excellent advantages when it comes to technology. It has a highly educated population, a large portion of who speak English. A sizable number of trained engineers who have emigrated from Russia are employed in the high-tech center. Because there are so many of them, Israel is able to pay its highly skilled workers much less than they would earn in the US.

A further advantage is that because Israel is relatively small in size, and somewhat removed from other countries, the environment is beneficial for communication and distributing new concepts. In addition to its geographical location, Israel has had to be somewhat independent due to conflict with the nearby Arab territories. This discord has actually proven somewhat advantageous to Israel's technology industry. Because of the elevated defense since 911, Israel has attracted attention for its businesses involving security functions.

The state of war that has plagued Israel since its foundation is a curse that paradoxically has produced beneficial side affects: for example, the development of its own highly sophisticated computerized defense system. In addition, Israel also has free trade agreements with the US and Europe. As a result, Israel boasts numerous networking and high technology companies.

There are Israeli companies in the United States all of which share common characteristics. First, the technology is developed in Israel, but the country itself is too small to sustain all of the companies. They, therefore, come to the U.S. to market their products and grow their companies. Much of the Israeli technology is strongest in niches such as wireless and fiber optics that fit its military priorities. Promising young recruits in Israel are given the opportunity to attend educational institutions that concentrate on these programs.

Israel has the best institutions of education and scientific research in the Middle East. These include the Technion-Israel Institute of Technology in Haifa, the Weitzmann Institute of Science in Rehovot, The Hebrew University in Jerusalem, and Ben Gurion University in the Negev. Two of the renowned world centers of high technology learning and application are the Technion-Israel Institute and the Weitzmann Institute. Scientists from all over the world come to study at these institutions. There are also numerous engineering students who are on long waiting lists for entrance into these prestigious establishments. Only a small number of places in the world offer a doctorate in phototonics, the transmission of light waves for optical and wireless use. The Weitzmann Institute of Science is one of them. The Technion Institute of Technology and the Hebrew University both have strong programs in optics and biotechnology.

After the 1993 Yom Kippur War, the Israel air force became the world leader in drones, or Unmanned Aerial Vehicles. Because so many pilots were lost in reconnaissance missions, it became a national priority to develop UAV's. There is currently a private Israeli-American company, Pioneer UAV, which is a 50-50 joint venture. Half the work is done in Israel, half in the United States. The construction of the unmanned aerial vehicles is literally divided between the two countries. The tail and wings are made in Israel, fuselage and assembly in the U.S.

CheckPoint Software Technologies is also a firm that makes Internet security software. It began as a result of the Israeli military's need for protection from "cyber-terrorists". This Israeli company had developed firewall programming.[9]

There are countless similar companies to CheckPoint Software. Because of the Internet's widespread use, a need for a certain level of organization has arisen.

> "MATIMOP - the Israeli Industry Center for Research and Development is a public non-profit organization, founded by the three major associations of manufacturers in Israel. Functioning as the interface between Israeli companies and their international counterparts, to promote joint developments of advanced technologies, MATIMOP encourages participation in the many international programs for bi-lateral and multilateral cooperation in Industrial Research and Development, signed and funded by the Office of the Chief Scientist (OCS) of the Ministry of Industry and Trade." [10]

This company heads a number of on-going projects relating to the technological progress of Israel. Some current projects are the pursuit of an Expert System for Equipment Diagnosis and Maintenance Support, an Environment Monitoring and Control System, and a Security System of a wireless form. Additionally ensuing is an intriguing collaboration on the Internet between many different territories. This group effort is known as the Geant Project and was established by 26 National Research and Education Networks representing 30 European countries. Israel along with several other non-European nations is now a full member of this endeavor, providing funding for the four-year mission through the Israeli InterUniversity Computation Center. By February of 2002, Israel maintained one of the fastest connections to the GEANT network.[11] This significant progress could be seen as a possible threat to the five major and nearly seventy smaller Internet Service Providers. One of these chief networks is MACHBA, an organization that works in the area of telecommunications (communications and computers), to enhance cooperation and mutual assistance among its member institutions, and between those members and other research institutes and organizations who share these interests and who are engaged in research and teaching on the university level.[12] Cyprus is yet another of these companies. CyNet is Cyprus' National Research and Education Network. It provides a network infrastructure for the Cypriot Research and Education Community. CyNet connects universities and research institutions.[13]

Israel is now second only to the United States in computer related start-up companies. Hank Nussbacher, one of the men accountable for the creation of the Israeli Internet, states that "Israel is about three years behind the US concerning the technological uses of computer networks. With the development of commercial services in the US, people order and send flowers, make plane reservations and complete other day-to-day tasks on their computers, such as reading major newspapers, The Jerusalem Post is among them. The fastest growth areas in the US are commercial networks, which businesspeople now use extensively to explore potential markets, seek out and communicate with their customers, and even to close deals"[14]

A successful company, Mirabilis, started by three young Israeli software designers and their venture capitalist backer was recently sold to America Online for 387 million dollars. Mirabilis is an Internet communications tool that allows users to talk and exchange computer files while online. Israel has developed a knowledge economy. Furthermore, several high-tech U.S. giants, IBM, Motorola, and Compaq, among others, have subsidized branches for sales and technical support in Israel.

It is important to question what it means for the future of the Middle East that while Syria, for example, is still debating whether to get on the Internet, Israel is already designing the next generation of the Internet. Moreover, an Israeli company near Tiberias is the only one in the world that makes a Key Ethernet-switching chip. Computer software developed in Israel accounts for more than half of the most popular personal computer downloads in the world. Israeli companies also dominate a key technology sector in making on-line tools for Internet security. It can even be said that Israel has its own Silicon Valley.

Notes

1. Friedman, Tom. *Lexus and Olive Tree*, Farrar Straus & Giroux, 1999, p. 9.
2. Foldvary, Fred E. and Klein, Daniel B. *The Half-Life of Policy Rationales: How New Technology Affects Old Policy Issues.* Internet: http://www.nyupress.org/webchapters/0814747760intro.pdf

3. Piore, Dr. Emanuel R. *The Computer Revolution*. New York: The City College, 1966.

4. Ibid.

5. Takaji, Yuji. *The Internet Age: Japan's Challenge to E-Business*. The University of Texas at Austin: Austin, TX, May 2003.

6. Ibid.

7. Grossman, Marc. *Technology and Diplomacy in the 21st Century*. Washington, DC: Sept. 2001

8. Herman, Larry. *The Hype and the Myth: The Future Isn't What It Used to Be*. Internet: http://www1.worldbank.org/publicsector/egov/herman_kpmg.pdf. June 2001.

9. Pash, Barbara. "30 Israeli businesses have come to Maryland seeking the Promised Land," *Baltimore Jewish Times*, July 27, 2001, p. 46.

10. Matimop: Israeli Industry Centure for R&D http://www2.matimop.org.il/1/general/about.asp, June 3, 2005.

11. http://www.internet-2.org.il/faq.html#4

12. Machba The Inter-university Computation Center http://www.machba.ac.il/html/framesets_eng/about_iucc.html

13. Cyprus Research and Academic Network http://www.cynet.ac.cy/english/CyNet_Home.htm, April, 2004.

14. Kaplan Sommer, Allison, "Hooked on Internet", *The Jerusalem Post*, May 6, 1994, p. 12.

CHAPTER 8

Conclusion

Technological changes take place in societies, affecting both the social and personal lives of their members in myriad ways. These are often taken for granted and thought of as almost natural. A case in point for the United States was the introduction of the automobile. The substitution of a motor-drawn for a horse-drawn vehicle was an invention of such far-reaching dimensions that it led to an automotive culture. Values and patterns of behavior associated with all of society's major institutions from the economy to the family were consequently affected and changed.[1]

The integration of new technologies into a society is not automatic. The social structure can provide conditions of conduciveness or impediments to the process of technological innovation. However, once technological change is initiated, it leads to consequences often quite unanticipated. This has been demonstrated in the case studies we have explored. Each of the case studies examined exemplifies the interactive dynamics of the relationships between a society's social structure, values and technological development as well as the impact of such technology upon all of a society's major institutions.

The Mbuti Pygmies who inhabit the Ituri rain forest in present-day Zaire and the Alaskan Eskimos illustrate the use of the simplest technology: tools developed for a hunting and foraging mode of existence. This earliest known mode of human existence provided a subsistence economy geared to sharing and cooperative efforts. Such societies depend for their existence upon animals and plants supplied by the natural environment. Food is not produced. All members of the group have equal access to the means of production (simple tools such as arrows, spears, harpoons, knives and lances). Materials for these tools are, moreover, available in the natural setting. Cooperation and sharing are essential for both group and individual survival. These eliminate social distinctions, for no one owns anything that others do not possess. Indeed, there are societal pressures that discourage the accumulation of individual material possessions such that anyone unwilling to share is ridiculed and frowned upon. Sharing is an important value necessitated by the demands of the environment, the group's limited technology and the available food supply. Within such a structure there is generalized reciprocity. Even when an individual hunts alone, others share in the consumption of his quarry. As a result, egalitarian values predominate within such a culture.

In these societies, the constant search for food requires some degree of nomadism and mobility, which in turn limits population size. For the Mbuti Pgymies, the division of labor, which requires male youths to spend a great deal of time with adult male hunters while female youths are off with the women gatherers, reduces the amount of heterosexual intimacy. As an added natural form of birth control, there is the custom that following birth the mother breastfeeds her child for three years, during which time she is not expected to have sexual relations with her husband.[2] A significant parallel exists between hunting and gathering societies and industrial societies in that in both societies geographical mobility and smaller family size are functional to the pursuit of valued goals.

For both Pygmies and Eskimos, marriage relationships are generally monogamous. Pygmies live in bands consisting of several nuclear families united by kinship; Eskimos live in households consisting of several nuclear families. These arrangements are functional for survival, since they foster cooperation and interdependence.

In hunting and gathering societies, economic activity revolves around obtaining and distributing food, which is a valuable commodity. Since there is no surplus from production that can be hoarded, there is little concentration of power. The tools used for obtaining food are available to all, and sharing of food, which must be constantly replenished, is expected. Therefore, ownership of possessions does not confer special privileges, power or rewards. Only the skills achieved in hunting bring esteem and prestige. All people who demonstrate these skills have influence and prestige.

Experience and age are also valued. There is a minimum of ascribed status (status given at birth), for neither achieved skills nor possessions in the form of land or other worldly goods can be passed on to heirs. There is a simple division of labor in work activities based upon sex roles. The social structure in not highly differentiated. This contributes to a lack of hierarchical and stratified relationships in the societies as a whole, making them egalitarian, since their subsistence economies depend upon generalized reciprocity and recognition of the interdependence of the individuals on their groups.

In the Pygmy society, both men and women participate in hunting and food-gathering activities. This contributes to maintaining equality between the sexes. For the Eskimos, on the other hand, hunting is primarily a male activity, possibly because of the harsh physical environment with which they must cope. Consequently, men are accorded higher status than women; however, women's work is not totally devoid of economic significance. In addition to child care, Eskimo women prepare the food and sew the skins, chores that are equally important for individual and group survival. Therefore, both in Eskimo and Pygmy society, marriage is an economic partnership dedicated to the survival of its members. Cooperative efforts and conformity are encouraged in these cultures because everyone is dependent on the group for survival. Nonetheless, achievement and entrepreneurship are highly valued in a hunting and foraging subsistence economy because it is essential to have some adventurous, independent adults who will take the initiative in finding and securing food. Because survival requires inititative, such behavior aids in finding and securing food and is therfore recognized and admired.

Although successful hunting may lead to an abundant food supply, technology is used neither to harness nor control the physical environment of either culture. Instead, technology is adapted to the local, natural environment. The use of simple tools allows the population to live in balance with nature. Thus, the notion of technological progress does not exist. The Pygmies, for example, revere the forest as their protector and the source of their bounty. For them it takes on religious and sacred dimensions. Nevertheless, for over thousands of years the Pygmies have engaged in reciprocal exchange relations with tribal villagers whose settlements border the rain forest, trading surplus meat for metal-edged spears crafted by the tribal agriculturists. In recent years, however, in an effort to manage its lands and resources, the government of Zaire has attempted to move the Pygmies out of the forest. This policy is undermining the traditional economic and social relations of the Pygmies and the tribal villagers, leading as well to the exploitation of the rain forest for commercial purpose.

In recent years, the traditional cultures of both the Pygmies and the Eskimos have been under assault through exposure to outside forces and the introduction of more sophisticated technology. As the metal-edged spear has affected the hunting methods of the Pygmies, the introduction of the rifle as a replacement for the bow and arrow reduced the need for dependence and sharing among the Eskimos. Hunting has now become a solitary activity. The rifle has affected the subsistence economy and has also led to the excessive exploitation of the arctic environment by reducing the number of land and sea animals that dwell there.

Traditional whale hunting was dramatic and dangerous. The requisite bravery, courage and skill of the hunter were important sources of pride for the entire community. Now the traditional harpoons have been replaced by whale bombs and dart guns, thereby lessening the courage and skill necessary for the hunt. On the one hand, the introduction of commercial whaling has been made possible by the introduction of modern technology. On the other hand, whaling is now limited to those who have economic capital. No longer is the hunting crew composed exclusively of extended family members.

The changes and the resulting exposure to Western materialism and values have led some Eskimos to seek employment in the oil and natural gas fields or in the installation of government radar systems. Increasingly, Eskimos live in two competing worlds.[3]

Firearms, outboard motors, snowmobiles and the introduction of a cash economy have taken their toll on the native Eskimo culture. Consequently, their traditional way of life—which although meager was nevertheless acceptable—has now been lost for many. As a result, some young people have lost the ability to live off the land and in the future may be forced to live on welfare, becoming strangers in the land of their forefathers.[4] The transition to a new technology has led the Eskimos to subject themselves to many unforseen and unintended consequences of what had originally been perceived as short-term benefits. The Eskimos represent an outstanding example of how an entire way of life and technology were adapted to a harsh environment. Different animals were hunted seasonally, and weapons were made from the bone and skins of these animals. There were no excess possessions, just the basic utilitarian needs of a subsistence economy were provided.

The simple technologies required for hunting and gathering societies, as exemplified by the Pygmies and the Eskimos, reveal some of the basic egalitarian human patterns that existed and were functional before the advent of more complex modern technologies. These newer technologies ushered in changes in the social structures of these societies, resulting in class stratification, private ownership of the mesans of production and political complexity.

The East African villages of Kenya and the agrarian villages of the Punjab region of India illustrate the use of cultivation technologies. When tools were developed that allowed for crops to be planted, animals to be domesticated and surplus to be produced, people began to live in permanent settlements. As a consequence, there were fundamental changes in their social structures. An individual's or a group's relationship to the land became crucial for determining placement in society.

Kenya, with its simple hoe and digging stick agricultural system, developed in a different fashion from the Punjab, with its more complex plow technology. Over time, both societies had to adapt to technological changes introduced by outside colonial forces. Traditionally, in Kenya, most people were engaged in shifting cultivation, a technique where land is slashed, burned and rotated. This method was suitable for both the physical environment and the state of technological development. The hoe and digging stick limited the size and scale of production. Production was on a small scale and was primarily for consumption. The social system was relatively egalitarian. Land was collectively used even though it was individually owned.

In East Africa, moreover, the lack of the use of animal power and the absence of wheel technology had an impact on the social structure. Families only grew enough crops to last a season. A feudal landowning class did not develop. Families used a piece of land for a few years and then moved on to new plots. Clearing of the fields was typically men's work; planting and weeding were viewed as women's responsibility.

Since women and children were an important source of labor in Kenya, polygamous marriages were encouraged. Multiple wives enabled a man to increase his assets. Having several wives and thus many children prevented the need for hired labor. This system permitted men to work less. One of the strongest appeals of polygamy for men was its economic benefits. A man with several wives could cultivate more land, produce more food for his household and achieve higher status as a result of increased production.[5] Women, in turn, often enjoyed considerable freedom of movement and some economic independence from the sale of their own crops. They were an integral part of the work process, had a stake in the economy and therefore acquired a voice in decision making.

In contrast, women were removed from the agricultural process in the more labor-intensive, sophisticated plow cultivation used in the Indian villages of the Punjab. As a consequence, the position of women with respect to men was subordinate to that of African women. In the Punjab, women were valued as mothers; in East Africa, they were valued as both workers and mothers. Women are more economically productive in hoe and digging-stick cultures, which explains why such societies are more egalitarian in their relationships between the sexes than plow-cultivation societies.

The sexual division of labor in farming is related to the complexity of the technology. Whenever the plow is used, it is employed almost entirely by men, who thus dominate the productive roles. In addition, in these societies all large livestock, such as horses, cattle, and camels, are relegated almost exclusively to male hands.[6] Where women participate in the cultivation process, husbands most often pay a bride price. In such societies, because wives are involved in the cultivation work, they are viewed as a souce of income and a form of investment. By contrast, where women do not participate in the economic activities, wives most often pay a husband price. The economic aspect of the marriage relationship is apparent in these arrangements.

All forms of cultivation are more labor-intensive than hunting and gathering. However, African hoe farming requires less preparation of land than plow cultivation and does not require the use of draft animals. Land can be used and is periodically rotated and allowed to remain fallow while new land is then cultivated. In the Punjab, the use of plow cultivation increased the productive potential of the land. Traditionally, however, given the social system, the surplus was pocketed by officials of the state, leaving little incentive for peasants to produce more than the minimum required for subsistence. Only after the introduction of an outside force in the guise of colonialism did changes in land ownership and cultivation techniques take place in the Punjab. At that point, land itself became an important commodity and agriculture became a lucrative business for those who owned land.

Both in Kenya and the Punjab, the impact of colonialism led to the development of cash crops, mechanized farming and private land ownership. In Kenya, the egalitarian system was upset by a landowning class that was able to sell or dispose of their land for profit. A dual farming system was created. Wealthy farmers used the new technology of the ox and plow, and, later, the tractor; poorer farmers continued to use the hoe and digging stick. Traditionally demarcated lands gave way to individually owned farms. Collective land for farming was no longer available for all. As a result of these innovations, some men were forced to work as agricultural laborers away from home. Consequently, women were left to shoulder the entire responsibility for farming chores, which had once been men's work. Due to changes in technology, the entire organization of work as well as the family structure was altered.

Children lost their former economic value and monogomous marriages became preferred. Familial values were replaced by individualistic ideals.

Under British rule in the Punjab, land for the first time became a valuable commodity. An unanticipated consequence of the newly valued land was the strengthened position of the moneylender in the village economy. Many farmers became indebted to moneylenders for loans to pay their taxes. These debts served as a disincentive for farmers to innovate. Later, when social incentives were provided by the government (such as liberal credit arrangements), the power of the moneylender was undermined and mechanization resulted. Mechanized farming destroyed the traditional equilibrium between the castes. Once mechanized technology was accepted, other changes occurred. Wealthy farmers who used tractors benefited enormously from the use of this improved method of production.

In the Punjab, the use of modern farm equipment promoted the adoption of new high-yielding varieties of seeds that ushered in the green revolution. The willingness to adopt new agricultural technologies only occurred when some farmers became proprietors and increased their land holdings. Only among the farmers who prospered was there motivation to innovate. Mechanization was not welcomed until a farm family achieved a level of economic well-being that safely allowed for departure from the traditional means of cultivation.

Advanced technology brought prosperity to some agriculturists, while, at the same time, others remained poor. Mechanization, land consolidation and irrigation facilities were needed to benefit from the use of the new high-yielding varieties of seed. The new seed technology was adopted only where such facilities were already available. As a result of mechanization and the use of this more efficient seed, the traditional caste ties were disturbed, increasing disparities between the rich and the poor. Case studies of both the Punjab and Kenya demonstrate the significance of structural and value dimensions in either inhibiting or promoting technological change. Both societies have remained in an agricultural mode, but have partially shifted to the use of a new technology, which uses mechanized rather than human and animal power.

The case study of Lowell, Massachusetts, illustrates a technology based upon new energy sources that resulted in a shift to machine from human and animal power. By means of a machine technology, the farm community of East Chelmsford was transformed into a modern industrial center. Work shifted from the household to a centralized workplace. An elaborate class system developed, as did a new political ideology: industrial capitalism. Lowell demonstrates the drastic changes that occur as a society shifts from one energy source to another, as well as when it moves from one mode of production to another, namely, from farming to manufacturing.

The desire for progress and greater productivity provided the impetus for men such as Francis Cabot Lowell to develop power-driven machinery. They believed that the conditions of the new world and the values of the republican society would make possible a new utopian form of industrialization. And they wished, thereby, to avoid the social ills that had plagued Manchester, England.

The detailed division of labor demanded by the use of the new technology led to the replacement of skilled craftsmen by semiskilled workers. Within the factory system, no individual would ever again produce a total product. The first industrial workers were, for the most part, young unmarried farm women; later, they were replaced by an immigrant work force. The new technology required changes in the rhythms and patterns of work, demanding long hours, disciplined work schedules and vertical stratification in the workplace.

Over time, the Yankee women, the first mill workers in Lowell, found these conditions intolerable and incompatible with their republican values of self-reliance and control over their activities. Poor and desperate for work, the new immigrants proved more willing to accept their lot as mere extensions of the machine. The urban ills of Manchester, England, were replicated in spite of the utopian ideals of Lowell and his associates.

The attempt in Lowell to unite the values of the former agrarian society with those of mass-production failed. The emphasis of capitalism on profit, private control of the means of production, private ownership and mass production superseded all other value premises. Profound, unanticipated changes in the quality of life were the by-products of the shift to work for wages in a workplace separated from the family farm. The concept of America as an agrarian Eden could not be reconciled with the machine technology. New technology inevitably leaves its own imprint, which in turn often results in value clashes. The transition to an industrial society entailed a severe restructuring of work routines that often was at odds with the values associated with a preindustrial way of life.

Certain crucial factors are present in the use and development of the technologies represented in all the case studies. Relationship to the means of production influences the structure of social relationships between individuals and groups and affects a society's orientation toward sharing and cooperative efforts. Materials for tools are available to hunters and gatherers in nature. Resources are renewable and the technologies are so simple that an accumulated surplus is not possible. In such a milieu, group sharing is the accepted norm and status is based on achieved skills, as opposed to status given at birth. There are social mechanisms that prevent individuals from dominating others as well as leveling mechanisms that emphasize egalitarian values. Religious beliefs are intertwined with the sacredness of the physical environment and the desire not to exploit nature. Such abuse would, of course, have serious consequences for food production.

In contrast to hunters and gatherers, agriculturists and industrialists produce surpluses. They also accumulate capital and exploit resources. Private ownership of land and machinery results in control by a few and leads to concentration of power.[7] Private interests often supersede the welfare of the group as a whole.

The world view, the implicit beliefs, values and assumptions about the nature of reality, is interrelated with the form of technology that a society maintains.[8] The world view of users of renewable energy sources—hunters and gatherers and cultivators—is fundamentally different from that of nonrenewable, energy-based industrial societies. The world view of a culture reflects its relationship to the natural environment, which provides its energy base as well as its religious and social values. It is functional for agriculturists to have their children learn obedience, responsibility and conformity to social rules; the crop must be planted and harvested. These chores require planning and adherence to a disciplined work schedule. Hunters are more spontaneous; lack of success one day may be made up the next.[9]

More complex technology leads to an irreversible path of increased division of labor and other forms of social differentiation. As a more sophisticated technology develops, there is less social interdependence but more technological interdependence. Greater specialization in the work process coincides with more stress on individualism. With industrialization, a decrease in worker autonomy is accompanied by the separation of work from family and community. Means become available for pursuit of individual prosperity, interests and goals.

Technology, by altering the environment in which relationships take place, obligates people to search for new values. One key element of the industrial society, in contrast to the societies of hunters, gatherers and agriculturists, is the expectation of progress. A technology that can provide mass production is viewed as a force that will liberate humankind from its resignation to a fate imposed by a world view of scarcity. The satisfaction of basic economic needs will lead to the resolution of social problems.

Over time, the organization of work required for operating the machine technology became the model for behavior associated with all other areas of life. Increased output required efficiency and disciplined performance. A market economy developed that emphasized competition and the unequal distribution of goods. Both the workplace and the community at large became highly stratified. In Lowell, people were stimulated to consume and produce wealth with the promise that they could keep what they produced as a reward for their efforts.

Each case study primarily has described a particular kind of technology and its influence on a society as well as the interplay of a society's values on the development of the technology. It has been shown that neither technology nor society is autonomous. Societies do not passively accommodate or receive new technology. Their values and social format promote or restrict its development. There is often a cultural lag when new technology is introduced. For a period of time, values appropriate to old patterns of work and production generally persist.

In Lowell, agrarian values proved inconsistent with the work relations required for machine technology. The machine is characterized by automatic action. This circumstance alters the workers' relationship to it. By contrast, the use of tools is controlled by the worker. Technological change does inevitably alter the character of work and has an effect on the structure and social organization of all institutions. Tools developed for cultivation of land led to permanent settlements and the creation of surplus. Agriculturalists were less sharing and more power-oriented than hunters and gathers. Relationship to land ownership promoted status distinctions. Ascribed status became a factor in a stratified hierarchy.

There are always unanticipated consequences of modes of technology. Under the pressure of the outside force of colonialism, mechanized farming was introduced in both East Africa and the Punjab. Among the unexpected results were changes in land ownership, a shift from subsistence to cash crops and a dual system of farming. The introduction of new technology always creates a disequilibrium in social structure and values. A new technology can result in accidental or deliberate change.[10]

A deliberately planned change in Lowell was the decision to utilize a work force comprised of young, unmarried farmwomen. Industrialization always requires advanced planning for the management and coordination of a large work force. Unlike an agricultural society, in which farmers could grow crops wherever land was available, the industrialized society has to provide a centralized workplace.

Computers provide the connecting links in the oft-fragmented world in which we live. Extending from the workplace to the home, our generation is reliant on the computer for communication and as a source of instant information. Originally, many worried, however, about the cultural effects of computer technology. As the computer revolution began in the late 1970's and early 1980's, no one knew exactly what this impressive new machine would do to the world. And, even as the revolution was dawning, the computer was quite controversial. Social scientist were concerned with the possible problems the computer might create, saying it would increasingly isolate people.

While social arguments against the computer were common, economic arguments were even more widespread. Numerous economists predicted that computers would take jobs from millions of workers. Critics, like Shoshana Zuboff, worried that computers would become an overpowering force, and that the rise of computerized technology would decrease the necessity for human labor. Although many jobs today are accomplished by computers, we don not see the enormous unemployment rate or the general social dissatisfaction predicted by critics. Instead we have "work" take on new definitions. The workplace has been renovated by technology. Computers now play a major role in every economic sector, making it easier for large companies to globalize and connect skills to income in new ways.

This incredible technological change and its effect on the nature of work can be seen all over the world. In Lowell, Massachusetts, as in many other cities, computer technology brought in a new immigrant work force. Rather than work in mills, as immigrants did in the early 20th century, today they assemble computer parts. While the United States and cities like Lowell have experienced significant change with rise of computer technology, Israel is probably one of the best examples of the importance of computers in modern society. Computer technology and Israel's willingness and ability to expand upon it have made the small nation a global high-tech center. The San Francisco Chronicle article of April 4, 2002 explains Israel's technological vitality as and its newfound role in the global economy. According to the article, the violence in the Middle East is hitting closer to home for many Americans as more and more high tech firms move into Israel and therefore have a stake in the nation. A major concern, therefore, in the Israeli-Palestinian conflict is the decrease in venture capital in times of significant violence. Computer technology is an important concern in every area of life in Israel and has made the nation a concern for every capitalist country on the globe.

The benefits of the Computer Revolution have brought about profound change in the world, as we know it today. Traditional borders have been shattered in favor of more and more global connection and trade. Some philosophers have compared the effect of the invention of the computer to the effect of the invention of writing, as a new movement of self-expression and communication, providing a new way to connect with potentially everyone else on the planet. Due to the computer Revolution, individuals have a greater sense of independence and liberation.

Although controversial at the start, the educational, economic, and personal benefits of the computer proved most critics of the technology wrong. Computer power exploded in the 1980's and 1990's and as prices dropped, the technology became accessible to everyone. The Internet connected societies worldwide and became an economic force to be reckoned with. In his book, *The Computer Revolution: An Economic Perspective*, Daniel Sichel says that the computer revolution is a "key factor in economic growth and productivity . . . Business Weekly . . . told its readers that 'the productivity surge of the [late nineties] may reflect the efforts of U.S. companies to finally take full advantage of the huge sums they've spent purchasing information technology". Computers have met social and economic standards to become the most important and powerful tools of our age.

The power of the computer as a vehicle for social change is particularly evident in the case of Israel. A nation once based on agriculture; Israel has become a leader in world technology. Israel has seen its economy shift as industry moved into high-tech ventures. Today this small nation has software sold all over the world, and has exports to numerous overseas companies. Politically, Israel uses it technology to its advantage, developing advanced weaponry and working to remain an independent nation. Socially, Israel has developed strong education and has come to be the leader in scientific research in the Middle East. The computer has brought about significant change in Israel, shifting the country's economy, government and security, and society.

In hunting and gathering societies, technology requires a minimum division of labor based only on gender. Any man or woman can normally do what any other person of their sex could do. Industrialization has resulted in greater specificity of job-related tasks, in turn leading to other forms of social differentiation. Industrialization entails a severe restructuring of all institutions as well as emphasis on greater productivity and economic growth. Agriculturists, by contrast, are more often conservative and resistant to change; their highest goal is to protect their land. Hunters and gatherers, though willing to take risks, persist in the use of an unchanging technology unless confronted by outside forces.

Once a new technology is introduced, it has its own ethos and, accordingly demands new forms of work relationships and orientation to the means of production. Technology itself may be a catalyst for further technological innovations when there are recognized incentives for change. New technologies always result in discontinuities in a society's identity, values and social roles. Each of the types of technologically based societies we have examined have organized production, consumption, labor and leisure in specific socio-technological configurations. We have shown how the interrelationships of tools, machines and values influence the processes of work and social life in diverse forms of societies. We have also explored the capacity for technology and values to affect the perception and responses to reality as well as the identity of the members of each society and the way they manage to live together.[11]

We live in an age of fast-paced technological change. The historical case studies we have examined should lead to a more critical understanding of the extent and nature of the impact of technology on social institutions and the lives of individuals. Also highlighted has been how the value premises of each society, religious, political and social, have influenced the acceptance of technology. Insights gained as to the circumstances under which technological innovation is viewed as progressive and benefical and when it is viewed as harmful should challenge traditional preceptions of the interrelations of technology, society and values.

We have seen the hunting and gathering societies of the Pygmies and Eskimos undergoing technological change. These former cooperative subsistence societies are now altered. The question that remains unanswered is will the benefits of the material goods made possible for them by industrialization outweigh the disruption of their traditional way of life?

Technology is strongly intertwined with our future in ways we do not fully comprehend. However, we should critically use the knowledge that is available in assessing its impact on the shape of societies. It is our hope that what we have written will aid in this process.

Notes

1. James I. Flink, *The Car Culture*, Cambridge, Mass.: MIT Press, 1975, pp. 18-41.

2. Colin M. Turnbull, *The Mbuti Pygmies: Change and Adaptation*, New York: Holt, Rinehart, & Winston, 1983, pp. 56-57.

3. Andrew L. Yarrow, "Alaska Natives Try a Taste of Capitalism," *New York Times Magazine*, March 17, 1985.

4. Eric Alden Smith, "Inuit of the Canadian Eastern Arctic," *Cultural Survival Quarterly*, Vol. 8, Fall 1984, pp. 32-37.

5. Ester Boserup, *Women's Role in Economic Development*, New York: St. Martin's Press, 1970, pp. 35-51.

6. Jack Goody, *Production and Reproduction*, London: Cambridge University Press, 1976, p. 35.

7. Gerhard Lenski, *Power and Privilege: A Theory of Social Stratification*, New York: McGraw-Hill, 1966, pp. 84-85.

8. R. G. Collingwood, *Essay on Metaphysics*, London: Clarendon Press, 1940.

9. Richard A. Barrett, *Culture and Conduct*, Belmont, Calif.: Wadsworth Publishing Company, 1984, pp. 76-96.

10. Maurice N. Richter, Jr., *Technology and Social Complexity*, Albany: State University of New York Press, 1982.

11. Langdon Winner, "On Criticizing Technology," Public Policy, Vol. XX, No. 1, Winter 1972, p. 55.

Bibliography

Abbot, S. "Full Time Farmers and Week-End Wives: An analysis of Altering Conjugal Roles." *Journal of Marriage and Family*, Vol. 38, No. 1, February 1976.

Adams, R.M. *The Evolution of Urban Society*. Chicago: Aldine, 1962.

Ahmad, I. "The Green Revolution and Tractorization: Their Mutual Relations and Socio-economic Effects." *International Labor Review*, Vol. 224, No. 1, 1976.

Anthony, K.R.M. & Johnson, B.F. *Agricultural Change in Tropical Africa*. Ithaca: Cornell University Press, 1979.

Arligiani, Robert. "The Tablet and the Tool: An Essay on Technology and Human Values." *Journal for the Humanities and Technology*, 1982.

Aronson, Jonathan David, and Peter F. Cowhey. *When Countries Talk: International Trade in Telecommunications Services*. Cambridge: Harper & Row Publisher Inc., 1988.

Aviram, Aharon, Integrating ICT and Education in Israel for the Third Millennium, Ben Gurion University, (June 2000). http://www.21learn.org/acti/aharonict.html

Barrett, Richard. *Culture and Conflict*. California: Wadsworth Publishing Company, 1984.

Barrier, N.J. *The Punjab Alienation of Land Bill of 1900*. Duke University Program in Comparative Studies on Southern Asia, #2.

Baumann, Hermann. "The Division of Work According to Sex in African Hoe Culture." *Africa*, Vol. 1, 1928.

Bender, Thomas. *Community and Social Change in America*. Baltimore: Johns Hopkins University Press, 1978.

Cairncross, Frances. The Death of Distance: How the Communications Revolution Will Change Our Lives. Boston: Harvard Business School Press, 1997.

Dublin, T. *Women at Work*. New York: Columbia University Press, 1979.

Duffy, Kevin, *Children of the Forest*. New York: Dodd, Mead, and Company, 1984.

Dutt, R. *The Economic History of India*, Vol. II. New York: Augustus M. Kelly Publishers, 1969.

Eckes, Jr., Alfred E., and Thomas W. Zeiler. *Globalization and the American Century*. Cambridge: Cambridge University Press, 2003.

Eilam, Lior. *Land of Unlimited Opportunities*. 4 October 2000. 13 Oct. 2004. http://www.globes.co.il/serveen/globes/printWindow.asp?did=443671

Encyclopedia Britannica, Vol. 9.

Eno, A.L. (ed.) Cotton *was King*. New Hampshire Publishing Co., 1976.

Eshel, Tamir. *Coming Soon: Anything But Computers*. 27 July 1997. 13 Oct. 2004. http://www.globes.co.il/serveen/globes/printWindow.asp?did=355572

Evans, Eli N. *Telecommunications and the World Jewish Renewal*. 3 June 1997. 13 Oct. 2004.

http://www.revsonfoundation.org/publications_telecom.htm

Flink, J. *The Car Culture*. Cambridge: The MIT Press, 1975.

Foldvary, Fred E. and Klein, Daniel B. *The Half-Life of Policy Rationales: How New Technology Affects Old Policy Issues*. Internet: http://www.nyupress.org/webchapters/0814747760intro. pdf

Foner, Phillip (ed.) The *Factory Girls*. Chicago: University of Illinois Press, 1977.

Frankel, F.R. *India's Green Revolution: Economic Gains and Political Costs*. Princeton, N.J.: Princeton University Press, 1972.

Fried, M. *The Evolution of Political Society*. New York: Random House, 1967.

Friedl, E. *Women and Men*. New York: Holt, Rinehart & Winston, 1975.

Friedman, T. *The Lexus and the Olive Tree: Understanding Globalization*. Anchor; 1 Anchor edition, 2000.

Gendron, Bernard. *Technology and the Human Condition*. New York: St. Martin's Press, 1977.

George, S. *Feeding the Few: Corporate Control of Food*. Washington, D.C.: Institute for Policy Studies, 1981.

Ghosh, A.K. "Institutional Structure, Technological Change and Growth in Poor Agrarian Economies." *World Development*, Vol. 7, No. 4-5, 1979.

Gibbs, J.L. ed. *Peoples of Africa*. New York: Holt, Rinehart & Winston, Inc., 1965.

Giffen, N.M. *The Roles of Men and Women in Eskimo Culture*. Chicago: University of Chicago Press, 1930.

Gill, S.S. & Singhal, K.C. "Punjab Farmer's Agitation-Response to Developmental Crisis in Agriculture," *Economic and Political Weekly*, vol. 19, 1984, p. 1729.

Ginsberg, E. *Seminar on Technology and Social Change*. New York: Columbia University Press, 1964.

Goody, Jack. "Economy and Feudalism in Africa." *The Economic History Review*, Vol. XXII, No. 3, December 1969.

Goody, Jack. *Production and Reproduction*. Cambridge: Cambridge, 1976.

Graburn, N.H.H. *Eskimos Without Igloos*. Boston: Little, Brown & Co., 1969.

Gregory, Frances W. *Nathan Appleton: Merchant and Entrepreneur, 1779-1861*. Charlottesville: University Press of Virginia, 1975.

Griffin, K. *The Political Economy of Agrarian Change*. London: The Macmillan Press, 1974.

Grossman, Marc. *Technology and Diplomacy in the 21st Century*. Washington, DC: Sept. 2001

Gubser, N.J. *The Nunamint Eskimo Hunters of Caribou*. New Haven, Conn.: Yale University Press, 1965.

Guemple, Lee (ed.) *Alliance in Eskimo Society*. Seattle: University of Washington Press, 1972.

Gupta, D. "The Communalizing of Punjab, 1980-1985", *Economic and Political Weekly*, Vol. 20, 1985, p. 1189.

Gutman, H. "Work, Culture, Society in Industrializing America" in *American Historical Review*, 1973.

Hallet, J.P. *Pygmy Kitabu*. New York: Random House, 1973.

Harris, Marvin. *Cannibals and Kings*. New York: Vintage Books, 1978.

Harris, Marvin. *America Now*. New York: Simon and Schuster, 1981.

Harris, M. *Cows, Pigs, Wars and Witches*. New York: Vintage Books, 1974.

Hart, J.A. & Hart, T.B. "The Mbuti of Zaire." *Cultural Survival Quarterly*, Vol. 8, No. 3, Fall 1984.

Herman, Larry. *The Hype and the Myth: The Future Isn't What It Used to Be*. Internet: http://www1.worldbank.org/publicsector/egov/herman_kpmg.pdf. June 2001.

Herschman, P. *Punjabi Kinship and Marriage*. Delhi: Hindustan Publishing Corp., 1981.

Heston, Alan W., ed., Richard D. Lambert, ed., Lamberton, Donald M., ed. *The Annals of the American Academy of Political and Social Science: The Information Revolution*. Philadelphia: Stanford University Press, 1974.

Howard, Alfred, *The Agricultural Testament*, London: Oxford University Press, 1940.

Hunter, G. & Bottrall, A.F. *Serving the Small Farmer: Policy Changes in Indian Agriculture*. London: croon Helm, 1974.

Itzikowitz, Jacob. *First Israeli Computer Project Completion and Continuation*. 1995. 13 Oct. 2004. http://itzikowitz.20m.com/Page08.html

Jaaskelainen, Sakari. *Andere Lander, gleiche Revolution?* 1998. 13 Oct. 2004. http://viadrina.euv-frankfurt-o.de/~sk/wired/global_ref.html

Jodka, S.S. "Crisis of the 1980's and Changing Agenda of Punjab Studies", *Economic and Political Weekly*, February 8, 1997, p. 274.

Josephson, H. *The Golden Threads*. New York: Duell, Sloan and Pearce, 1949.

Kaplan, Irwin (ed.) *Zaire: A Country Study*. Washington, D.C.: Foreign Affairs Studies of the American University, 1979.

Kassan, J.F. *Civilizing the Machine*. New York: Penguin Books, 1976.

Kenngott, George. *The Record of a City*. New York: Macmillan, 1912.

Kenyatta, J. *Facing Mount Kenya*. London: Secker and Wraburg, 1953.

Klausner, S.Z. & Foulkes, E.F. *Eskimo Capitalists: Oil, Politics and Alcohol*. New Jersey: Allanheld Osmun Publishers, 1982.

Knight, Will. *Scientists teach computer to speak*, ZDNET (UK). April 16, 2001. http://zdnet.com.com/2100-11-529299.html?legacy=zdnn

Kolack, Shirley. "Lowell, an Immigrant City: The Old and the New", Roy Bryce-Laport (ed.) *Sourcebook on the New Immigration*. New Jersey: Transaction Books, 1980.

Landau, Efi. *Aptonix: Be Acquired or Perish*. 27 December 2000. 13 Oct. 2004. http://www.globes.co.il/serveEN/globes/docView.asp?did=459682&fid=942

Layton, E.T. (ed.) *Technology and Social Change in America*. New York: Harper & Row, 1973.

Leaf, M.J. *Information and Behavior in a Sikh Village*. Los Angeles: University of California Press, 1972.

Leaf, M.J. *Song of Hope: The Green Revolution in a Punjabi Village*. New Brunswick: Rutgers University Press, 1984.

Leaky, R.E. & Larin, R. *People of the Lake*. New York: Avon Books, 1978.

Lee, R.B. *The ! Kung San: Men, Women and Work in a Foraging Society*. New York: Cambridge University Press, 1979.

Lefaucheux, M.H. "The Contribution of Women to the Economic and Social Development of African Countries." *International Labor Review*, Vol. II, 1962.

Lenski, Gerhard. *Power and Privilege: A Theory of Stratification*. New York: McGraw Hill, 1965.

Lenski, Gerhard and Lenski, Jean. *Human Societies: An Introduction to Macrosociology* (4th Edition.) New York: McGraw Hill, 1982.

Lineberry, W.P. (ed.) *East Africa.* New York: H.W. Wilson Co., 1968.

Lipschultz, David. *Bombs Yes, But No Crash in Israel.* 21 May 2002. 13 Oct. 2004. http://www.cji.co.il/cji-n212.txt

Luckham, M.E. "The Early History of the Kenya Department of Agriculture." *East African Agricultural Journal*, Vol. XXV, No. 2, October 1959.

Machlis, Avi. Multimedia revolution may be bypassing Jewish world. http://www.jewishsf.com/bk010511/supp9a.shtml

Majumdar, D.N. & Madan, T.N. *An Introduction to Social Anthropology.* Bombay: Asia Publishing House, 1963.

Mamdami, M. *Myth of Population Control.* London: Monthly Review Press, 1972.

K. Marx, *A Contribution to the Critique of Political Economy*, Cambridge: Harvard University Press, 1970.

Mesthene, E. Technological Change: *Its Impact on Man and Society*, Cambridge: Harvard University Press, 1970.

Morgan, Lael. *And the Lord Provides: Alaskan Natives in a Year of Transition.* New York: Anchor Press/Doubleday, 1974.

Moore, B. *Social Origins of Dictatorship and Democracy.* Boston: Beacon Press, 1966.

Moore, W. et al (ed.) *Labor Commitment and Social Change in Developing Areas.* New York: Kraus International Publishers, 1960.

Moore, R.K. *Technology and Social Change.* Chicago: Quadrangle Books, 1972.

Nass, Gilad. *100, 000 ADSL Customers in Israel.* 12 August 2002. 13 Oct. 2004. http://www.cji.co.il/cji-n218.txt

Nelson, R.K. *Hunters of the Northern Ice.* Chicago: The University of Chicago Press, 1969.

Nelson, R.K. *Shadow of the Hunter.* Chicago: The University of Chicago Press, 1980.

Nisbet, R. *Man and Technics.* Tempe: Arizona University Press, 1956.

O'Connell, William Cardinal. *Reflections of Seventy Years.* Boston: Houghton, Mifflin, 1934.

Ogburn, William. *Our Culture and Social Change; Selected Papers.* Chicago: The University of Chicago Press, 1964.

Pacey, A. *The Maze of Ingenuity.* New York: Holmes & Meier, 1975.

Piore, Dr. Emanuel R. *The Computer Revolution.* New York: The City College, 1966.

Putnam, A.E. *Madami: My Eight Years of Adventure with Congo Pygmies.* New York: Prentice Hall, 1954.

Ray, D.J. *Ethnohistory in the Arctic: The Bering Strait Eskimo.* Ontario: The Limestone Press, 1983.

Reed, C. (ed.) *Origins of Agriculture.* The Hague: Mouton, 1975.

Richards, A. *Economic Development and Tribal Change.* Cambridge, 1952.

Richter, Maurice N., Jr. (ed.) *Technology and Social Complexity.* Albany: State University of New York, 1982.

Richter, Maurice N., Jr. *Society: A Macroscopic View.* Cambridge, Ma: Schenkman.

Roberts, W. & Kartar Singh, S.B.S. *A Textbook of Punjab Agriculture.* Lahore, 1951.

Robinson, Harriet. *Loom and Spindle or Life Among the Early Mill Girls*. Hawaii: Press Pacifica, 1976.

Rossi, A. et al (ed.) *The Family*. New York: W.W. Norton & Co., Inc., 1978.

Schellenberg, Kathryn, Ed. *Computers in Society*. (1996) Dushkin Publishing Group/Brown & Benchmark Publishers: Connecticut, Vol. 6

Shenker, Hillel. *The Revolution is here, in Israel? E-Learning in Israel*. 2001. 13 Oct. 2004. http://www.jafi.org.il/aliyah.dept/aliyon/aliyon10/10.html

Shiva, Vandana, *The Politics of the Green Revolution: Third World Agriculture, Ecology and Politics*, London: Zed Books, 1991, p. 26.

Siedman, A. *Comparative Development Strategies in East Africa*. Nairobi: East African Publishing House, 1972.

Siedman, A. *Planning for Development in SunSaharan Africa*. New York: Praeger Publishers, 1972.

Smelser, N. *Social Change in Industrial Revolution*. Chicago: The University of Chicago Press, 1959.

Sobelman, Ariel T. *An Information Revolution in the Middle East?* June 1998. 13 Oct. 2004. http://www.tau.ac.il/jcss/sa/v1n2p4_n.html

Social Implications of Industrialization and Urbanization in Africa South of Sahara. UNESCO, 1956.

Southall, A.W. (ed.) *Social Change in Modern Africa*. London: Oxford University Press, 1961.

Spenser, R.F. *The North Alaskan Eskimo*. Washington, D.C.: Smithsonian Institute Press, 1959.

Spicer, E.H. (ed.) *Human Problems in Technological Change*. New York: Russell Sage Foundation, 1952.

Stevensen, W.K. "Farmers of Punjab are India's Shining Example." *New York Times*, October 7, 1982.

Takaji, Yuji. *The Internet Age: Japan's Challenge to E-Business*. The University of Texas at Austin: Austin, TX, May 2003.

Teitelbaum, Sheldon. *Cellular Obsessions*. January 1997. 13 Oct. 2004. http://www.wired.com/wired/archive/5.001/ffisraeli.html

Tignor, R.L. *The Colonial Transformation of Kenya*. Princeton, N.J.: Princeton University Press, 1976.

The Lowell Offering, prepared by females employed in the mills, Lowell, 1840–1845, New York: Greenwood Press, 1970.

Thompson, E.P. *The Making of The English working Class*. New York: vintage Books, 1966.

Tully, M & Jacob, S. *Amritsar Mrs. Gandhi's Last Battle*, London: Jonathan Cape, 1985, p. 49.

Turnbull, C. *The Forest People*. New York: Simon and Schuster, 1962.

Turnbull, C. *Wayward Serrants*. New York: Natural History press, 1965.

Turnbull, C. *The Mbuti Pygmies: Change and Adaptation*. New York: Holt, Rinehart & Winston, 1983.

Ucko, P.G.W. et al (eds.) *Man, Settlement and Urbanism*. London: Duckworth: 1972.

Vesilird, P.J. "Hunters of the Lost Spirit." *National Geographic*, Vol., 163, No. 2, 1983.

Wallace, A. (ed.) *Men and Cultures*. Philadelphia: University of Pennsylvania Press, 1960.

Wally, Herbert. *Eskimos*. London: Collins Publishers Franklin Watts, Inc. 1976.

Ware, C. *The Early New England Cotton Manufacture*. New York: Russell & Russell, 1966.

Wells, F.A. & Warmington, W.A. *Studies in Industrialization*. London: 1962.

White, L. *The Evolution of Culture*. New York: McGraw Hill, 1959.

Zuboff, Shoshana. *In the Age of the Smart Machine*. New York: Basic Books Inc., Publishers, 1984

Widstrand, C.G. *Cooperatives and Rural Development in East Africa*. Uppsala: Scandinavian Institute of African Studies, 1970.

Winner, Landon. "On Criticizing Technology" *Public Policy*, Vol. XX, Winter 1972.

Yarrow, Andrew L. "Alaska's Natives Try a Taste of Capitalism." *Sunday Magazine, New York Times*, March 17, 1985

Index